A HISTORY OF RADIOECOLOGY

This book presents a history of radioecology, from World War II through to the critical years of the Cold War, finishing with a discussion of recent developments and future implications for the field.

Drawing on a vast array of primary sources, the book reviews, synthesizes and discusses the implications of the ecological research supported by the Atomic Energy Commission (AEC) of the United States government, from World War II to the early 1970s. This was a critical period in the history of ecology, characterized by a transition from the older, largely descriptive studies of communities of plants and animals to the modern form of the science involving functional studies of energy flow and mineral cycling in ecosystems. This transition was in large part due to the development of radioecology, which was a by-product of the Cold War and the need to understand and predict the consequences of a nuclear war that was planned but has never occurred. The book draws on important case studies, such as the Pacific Proving Grounds, the Nevada Test Site, El Verde in Puerto Rico, the Brookhaven National Laboratory and recent events such as the nuclear disasters at Chernobyl and Fukushima. By revisiting studies and archived information from the Cold War era, this book offers lessons from the history of radioecology to provide background and perspective for understanding possible present-day impacts from issues of radiation risks associated with nuclear power generation. This is a timely contribution given the rising potential of nuclear warfare among world powers, the periodic accidents that occur at nuclear power plants and society's long-term failure to effectively dispose of nuclear wastes that continue to accumulate.

This book will be of great interest to students and scholars of radioecology, environmental pollution, environmental technology, bioscience and environmental history.

Patrick C. Kangas is a systems ecologist with interests in the structure and function of novel ecosystems. From 1990 until his retirement in 2021, he was an Associate Professor in the Department of Environmental Science and Technology at the University of Maryland. He has conducted research in Puerto Rico, Brazil, and Belize and has published more than 100 papers, book chapters and contract reports on a variety of environmental subjects. He is the author of *Ecological Engineering: Principles and Practice* (2004).

Routledge Explorations in Environmental Studies

Art, Farming and Food for the Future
Transforming Agriculture
Barbara L. Benish and Nathalie Blanc

Sustainable Places
Addressing Social Inequality and Environmental Crisis
David Adamson, Lorena Axinte, Mark Lang and Terry Marsden

Urban Aerial Pesticide Spraying Campaigns
Government Disinformation, Industry Profits, and Public Harm
Manuel Vallée

Addicted to Growth
Societal Therapy for a Sustainable Wellbeing Future
Robert Costanza

Coastal Disaster Risk Management in Bangladesh
Vulnerability and Resilience
Edited by Mahbuba Nasreen, Khondoker Mokaddem Hossain and Mohammed Moniruzzaman Khan

A History of Radioecology
Patrick C. Kangas

For more information about this series, please visit: www.routledge.com/Routledge-Explorations-in-Environmental-Studies/book-series/REES

A HISTORY OF RADIOECOLOGY

Patrick C. Kangas

Routledge
Taylor & Francis Group
LONDON AND NEW YORK

earthscan
from Routledge

Designed cover image: U.S. Atomic Energy Commission

First published 2023
by Routledge
4 Park Square, Milton Park, Abingdon, Oxon OX14 4RN

and by Routledge
605 Third Avenue, New York, NY 10158

Routledge is an imprint of the Taylor & Francis Group, an informa business

British Library Cataloguing-in-Publication Data
A catalogue record for this book is available from the British Library

Library of Congress Cataloging-in-Publication Data
Names: Kangas, Patrick C., author.
Title: A history of radioecology / Patrick C. Kangas.
Description: New York, NY : Routledge, 2023. | Includes bibliographical references and index.
Identifiers: LCCN 2022036728 (print) | LCCN 2022036729 (ebook) | ISBN 9781032257631 (hardback) | ISBN 9781032257655 (paperback) | ISBN 9781003284901 (ebook)
Subjects: LCSH: Radioecology—History. | Radioactive pollution of the atmosphere. | Radioactive pollution of water.
Classification: LCC QH543.5 .K36 2023 (print) | LCC QH543.5 (ebook) | DDC 577.27/7—dc23/eng/20220809
LC record available at https://lccn.loc.gov/2022036728
LC ebook record available at https://lccn.loc.gov/2022036729

ISBN: 978-1-032-25763-1 (hbk)
ISBN: 978-1-032-25765-5 (pbk)
ISBN: 978-1-003-28490-1 (ebk)

DOI: 10.4324/9781003284901

Typeset in Bembo
by Apex CoVantage, LLC

I dedicate this book to my parents, John and Helen Kangas, who raised me through the Cold War years, and to my brother, Michael Kangas, who helped start me on my career path in Ecology.

CONTENTS

PREFACE

This book concerns an episode in the history of ecology in the United States during the early years of the Cold War when ecologists were asked to assess the environmental impacts of nuclear weapons tests. This time period was marked by the geopolitical struggle between the free world of the Western powers and the Communist world of the Soviet Union, which lent an urgency to the ecologists' work. These ecologists studied nuclear bomb detonations and the fallout that was released. They injected radioisotopes into plants, soils and lakes and then tracked how these isotopes moved along food chains and cycled through the systems. They placed deadly sources of ionizing radiation in ecosystems and watched what happened. Many of their experiments would not be possible under today's rules and regulations for human safety and animal welfare. They had to interact with health physicists to avoid radiation hazard, and they sometimes had to deal with incomplete knowledge about what they were studying due to the secrecy of national security requirements associated with weapons testing. They responded to the challenges of Cold War science with a creative energy that helped transform the discipline of Ecology into its modern form. A number of the ecologists who carried out this research in Radioecology were role models for my own career in Ecology, so I've enjoyed highlighting their work. I hope this review of the literature captures some of the interesting and unique stories of this period in the history of ecology. I also hope this review shows modern readers that their research on ecological preadaptation and self-organization can enrich our current understanding of the novel ecosystems emerging during the Anthropocene.

ACKNOWLEDGMENTS

Special thanks are due to Jose-Luis Izursa, who drafted all of the figures in this book. His combination of artistic ability and mastery of computer software allowed him to capture the form and detail of any image I brought to him. I greatly appreciate access to materials from the E. P. Odum Collection at the University of Georgia Library, the H. T. Odum Collection at the University of Florida Library and the general collections at the University of Maryland Library. I also wish to thank Katie Stokes at Taylor & Francis Group for helping me see this project through to the end. Ms. Spandana P. B. and Apex CoVantage did an excellent job of editing the manuscript during the production stage which I appreciate. Finally, this work could not have been completed without the patience and support of my wife, Melissa. She assisted in many direct and indirect ways, especially at the end of the project when I relied on her literary knowledge and organizational skills to finish the job!

ACRONYMS AND ABBREVIATIONS

AEC	Atomic Energy Commission
AFL	Applied Fisheries Laboratory
BNL	Brookhaven National Laboratory
DOE	Department of Energy
ERDA	Energy Research and Development Administration
IBP	International Biological Program
NASA	National Aeronautics and Space Administration
NPT	Non-proliferation Treaty
NTS	Nevada Test Site
ORNL	Oak Ridge National Laboratory
SREL	Savannah River Ecology Laboratory

PART I

1

INTRODUCTION

The Discipline of Ecology and World War II

Before World War II Ecology was a small, new field characterized largely by descriptive studies focused on either plants or animals. It was an academic backwater in Biology departments where physiology, taxonomy and embryology were dominant. Ecological research mostly described the distribution of organisms along environmental gradients and the interrelations between organisms. Studies of succession, which defines how changes take place at a particular site over time after a disturbance, such as a hurricane or a fire, were important at this time. Most historians of science home in on the succession concept, and they have characterized pre–World War II Ecology as a philosophical struggle between those who believed organisms change over time as an organic whole (e.g., as a coordinated "superorganism") towards a final climatically determined state versus those who believed either that organisms change over time independently of other organisms (e.g., individualistically), or at least that the changes were less coordinated and could result in a greater variety of final stages rather than one climatically determined "climax" state. Overall, pre–World War II Ecology as a discipline was a kind of rigorous natural history mired in disputes about the distribution of organisms and their aggregations in space and time.

World War II fundamentally changed the state of the art of Ecology by pushing and pulling a significant group of ecologists in a different, new research direction. The cause of this change was the development of atomic bombs, which were used by the United States to end the war with Japan in 1945. The story of the development of the bomb (Rhodes 1986) and the use of two bombs on Japanese cities (Hiroshima and Nagasaki) in August 1945 is well known. Over 100,000 people were killed by the blasts and by the radiation disease that followed (Schull 1995). Even though this was a small fraction of the total number

DOI: 10.4324/9781003284901-2

of people killed during the whole of World War II, the death and destruction resulting from those two bombs (caused by splitting the atom to create enormous energy) was new and frightening, and it left a lasting impact on the world that continues today.

The atomic bomb was primarily developed to defeat Nazi Germany, which was feared to be developing its own bomb. A race took place in the Allied countries (in particular, the United States and Great Britain) to beat the Germans, but it turned out to be no contest. The Germans were unable to employ the necessary resources and make much progress at developing the bomb and they surrendered before the U.S. Manhattan Project completed its work on the weapon in mid-1945. However, Japan continued to fight after the German surrender and, ultimately, U.S. President Truman decided to use the new weapon to quickly end the war and avoid having to directly invade Japan.

The Cold War Begins

The atomic bomb (actually there were multiple designs for it) was the product of tremendous scientific collaboration by the physicists and engineers who designed and built it, along with the stupendous U.S. economy, which was capable of providing the support and the materials needed to produce the bomb. World War II had mobilized the people and industry of the United States on a grand scale to contribute to the war effort and the atomic bomb was an example of the military production needed to win. The freedom of the world was at stake and only through united sacrifice and effort could the war be won against the Axis powers of Germany, Japan and Italy. The wartime economy in the U.S. grew rapidly and efficiently to the point that at the end of the war the U.S. economy was providing half of the world's goods and services. As a demonstration of the magnitude and effectiveness of the U.S. war effort, the atomic bomb took only two and a half years to produce, from theory to the working prototypes that ended the war with Japan.

However, the bombs that brought World War II to a conclusion almost immediately started a new one: the Cold War between the U.S. and its Western allies and the Soviet Union and its satellite countries (McMahon 2003, Freedman 2001). Although allied with the West during the war with Germany, the Soviets vied for world domination after World War II. Their motivations were ideological differences (communism vs. capitalism) and security fears, as they were haunted by the destruction their country had suffered from Nazi Germany during World War II. The conflict started in Europe, where Winston Churchill declared in 1946 that an "iron curtain" had been set up between the West and the Soviet Bloc of the East. However, it quickly spread around the globe, where the Cold War sometimes turned into smaller "hot" wars, as in Korea and Vietnam. During the Cold War the stakes, in a geopolitical sense, were just as high as in World War II with freedom and free-market capitalism on one side and communism and totalitarian government on the other side.

The atomic bomb played a defining role in the Cold War. In the beginning it was the "trump card" of the U.S., the first nation to develop the technology, but the Soviet Union quickly developed its own atomic bomb and tested it in 1949. A true arms race began, with both sides accumulating atomic bombs (both fission and fusion weapons) and designing new delivery systems to get the bomb to its target (bomber airplanes, rockets such as intercontinental ballistic missiles and nuclear submarines). By the late 1960s the two sides—the West, led by the U.S., and the East, led by the Soviet Union—had achieved a state of mutually assured destruction in that each side had enough weapons to completely destroy the other, which created a kind of stalemate.

The Cold War continued until the late 1980s when new forms of diplomacy emerged between the East and West and the arms race ended. During the Cold War the world became much more complicated in terms of atomic energy with other countries developing nuclear weapons and with the development of commercial nuclear reactors for domestic power generation. New concerns about radiation came with threats of use of nuclear weapons by rogue states and terrorist groups and with the lack of sustainable disposal options for the nuclear wastes that continued to build up over time.

The Atomic Energy Commission and the Birth of Radioecology

Within the post–World War II context, the United States Congress in 1946 formed the Atomic Energy Commission (AEC) to replace the Manhattan Project in order to provide civilian control over the bomb and the development of nuclear technology. A primary role for the AEC was to oversee testing of new designs for the bomb. This was necessary because the arms race with the Soviet Union continually generated new weapons that required field testing as a preliminary step before arms production could occur. Thus, test sites were established and bombs were detonated in experimental trials, first in the atmosphere and on the ground surface and later belowground after concerns became apparent about fallout of radioactive debris from the detonations.

The environmental effects of nuclear testing were investigated from the first detonation at the Trinity site, in New Mexico, in 1945, but the early efforts were minimal. From the start the AEC focused to a greater extent on human health effects of radiation. However, the need for a deeper knowledge of the ecology of radiation and bomb effects became apparent after one of the tests in the Marshall Islands of the Pacific Ocean accidentally spread radioactive fallout over islands to which some native people had been relocated and over a Japanese fishing vessel (with the unfortunate name of *Fukuryu Maru—The Lucky Dragon*). This was the disastrous code-name Bravo test on March 1, 1954. Most of the people exposed to the fallout developed radiation sickness, which included hair loss, skin lesions and low blood count averages, and several died from the injuries. Not only were there severe impacts to humans from Bravo, but the tropical islands within the fallout shadow were contaminated with radiation to the degree that they were uninhabitable. Knowledge of the dangers of radioactive fallout from testing had been increasing before this time, but the Bravo accident was a trigger that caused the AEC to look at environmental impacts more broadly than it had previously (Kwa 1993, Rothschild 2013).

Fortuitously, in 1955 the AEC hired an ecologist from Ohio State University, John Wolfe, who later came to administer their new efforts at environmental assessment (Sprugel 1975). Eventually Wolfe developed a program that came to be known as Radioecology with extensive and well-funded research on radiation effects in the environment (Reichle and Auerbach 2003). Over the next 15 years the AEC funded research at levels never experienced before in ecology and this work transformed the discipline of Ecology from its pre–World War II state of a descriptive science of populations and communities mired in terminological disputes about succession into a new state of mechanistic and experimental science that studied energy flows and mineral cycling with computer modeling, along with the earlier descriptive approaches, all to help understand the effects on radiation. The key to the new AEC-funded ecology was the ecosystem concept and the methods used to study it.

Enter the Ecosystem Concept

The ecosystem concept was introduced by Arthur Tansley (though his colleague Roy Clapham actually may have come up with the term [Ayres 2012]) to clarify the pre–World War II terminology about succession and ecological superorganisms (Tansley 1935, van der Valk 2014). It was a vision of what was to come, but Tansley, though a very interesting person (Ayres 2012), was a traditional plant ecologist with little or no experience in the kinds of processes and data that would come to characterize the ecosystem. He was near the end of his career when he proposed the ecosystem concept and he did not pursue it. Instead he produced his magnum opus: *The Vegetation of the British Isles*, which was a landmark study in the style of pre–World War II type descriptive ecology.

Tansley's ecosystem concept had little impact on the discipline until Raymond Lindeman, a youthful ecologist who studied a small bog lake in Minnesota for his PhD research, captured its meaning, especially during a post-doctoral year at Yale University working with G. E. Hutchinson. With Hutchinson's guidance, Lindeman composed a quantitative model of energy flow for the system that was organized with trophic levels (Lindeman 1942). This was a transformational study that used traditional descriptive ecological data on populations of aquatic plants and animals that he had gathered in the lake to create a theoretical view of how the populations behaved and interacted to process energy in the system as a whole. Working in the field with only the help of his wife and with little financial support, Lindeman wasn't able to directly measure energy flows but he estimated them with assumptions and literature data. It was a heroic study within the history of ecology for the intellectual boost it gave the discipline (Cook 1977, Sterner 2012). Tragically, Lindeman's death in 1942 cut short a career with great potential and he didn't live long enough to see his final work published.

The accident with radioactive fallout from the Bravo test shot took place 12 years after Lindeman's death and it showed the need for large-scale, coordinated, interdisciplinary studies of the ecosystem to understand the effects of radioactive fallout and the other impacts of the atomic bomb on the environment. Bomb testing was releasing toxic radionuclides into the atmosphere that circulated around the globe creating pollution at a scale that had not been seen before and the science of Ecology lacked adequate tools and concepts to comprehend it. Society needed the ecosystem concept and scientists who could study it during the Cold War, and John Wolfe proved to be the right person to administer AEC-funded research that developed this kind of knowledge base. It was not Tansley's ecosystem nor even Lindeman's ecosystem that was needed but a new, operational approach to measuring and modeling the structure and function of Nature. The AEC-funded conception of the ecosystem emerged from different projects in different biomes carried out by groups of researchers led by individuals who lent their creativity to the concept. It was characterized, not just by new data on energy flow and mineral cycling, but also concerned how data on these processes could be combined with traditional descriptive data on populations and communities to arrive at a systems perspective of radiation impacts that was termed Radioecology. The systems models produced by radioecologists allowed cycles of radioactive fallout to be visualized and for radiation effects to be understood, but also they allowed new data on ecosystem function to be combined with the "old school" descriptive data that was still needed.

This book is about the Cold War–era Radioecology, which has largely been forgotten and never properly synthesized. Ecological studies that were related to the development of the atomic bomb from the end of World War II until the early 1970s are covered to illuminate the early, formative years of the ecosystem concept when it became operational. The ecosystem was a new level of organization that was inserted in the hierarchy of life above populations and communities, and AEC funding gave its study an inertia that can still be felt in today's discipline. A number of important studies from the Cold War era are covered in the book, illustrating the diversity in ecosystem types that was studied and the research leadership that characterized early Radioecology in the United States. Many of the people who led the projects became important figures in the history of ecology. They, along with the many young researchers who received their training in Cold War–era Radioecology, helped to transform the discipline of Ecology into the modern, post–World War II field that it is today. This kind of heritage is highlighted in each of the chapters of Part II of this book.

Additionally, comparisons of the Cold War–era Radioecology will be made with the recent state of the art of the discipline. During the Cold War the urgent concerns were about bomb effects during testing and, of course, to understand possible consequences of nuclear war. Present-day issues with radiation focus on consequences of nuclear reactor accidents and the disposal of nuclear wastes. Possible lessons that the Cold War–era research might offer present day Radioecology will be explored in Chapter 10.

The Bomb and the Ecosystem

When a nuclear bomb detonates, a large quantity of energy is released within a small volume of material. The immense power of a nuclear bomb, unlike a conventional bomb, comes from within the nucleus of the atom. The energy is released from the fission or splitting of heavy nuclei (e.g., uranium or plutonium) or from the fusion or combining of light nuclei (e.g., hydrogen). Bombs using nuclear fission are sometimes termed "atomic" and those using nuclear fusion are sometimes termed "hydrogen" or "thermonuclear". Nuclear explosions are characterized by extremely high temperatures and pressures that come from the nearly instantaneous releases of energy. The explosive energy of a nuclear bomb is measured by its yield compared with a kiloton (1000 tons) or a megaton (1,000,000 tons) of TNT. As noted by Eisenbud (1963):

> Approximately 50 per cent of the energy from a nuclear explosion is released in the form of blast effects, 35 per cent as thermal radiation and the remaining 15 per cent as ionizing radiation. . . . Of the ionizing radiation, one-third is prompt radiation produced within a few seconds after the detonation and two-thirds, or 10 per cent of the total energy of the explosion, is delayed ionizing radiation produced by decaying fission products and induced radionuclides.

Figure 1.1 illustrates these different components of the nuclear explosion. Only the radiation is unique to nuclear explosions, but all of them contribute to create a massive disturbance. The nuclear reactions of the bomb produce isotopes of different elements that release radioactive energy in the form of alpha, beta or gamma rays. Some of this radioactivity is spread away from the site of the detonation in the atmosphere, eventually leading to fallout (Klement 1965). Thus, detonation of a nuclear bomb causes effects close to the site of detonation and effects over a wider area.

Obviously, the detonation of a nuclear bomb over land or water will have major environmental impacts on an ecosystem. Ecologists differentiate between disturbance and stress, but both occur with a nuclear bomb detonation. To some extent the blast and fires caused by a bomb detonation represent disturbance at the landscape scale, but ecosystems have been exposed to these types of conditions from fires, hurricanes and volcanic eruptions for hundreds of millions of years (Turner et al. 1997). Disturbance triggers ecological succession, which will ultimately, over time, restore the ecosystem after the impact. The radiation from the

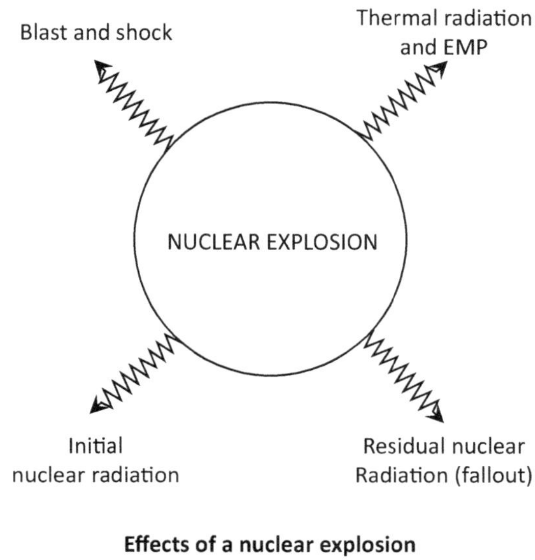

FIGURE 1.1 Overview diagram of the components of a nuclear explosion (from OTA 1979).

bomb, however, is different and it creates special stress at the scale of individual organisms. Life has evolved under natural conditions of low-level, "background" radioactivity, especially, for example, where uranium ore deposits occur. This fact has been utilized in bioprospecting to help locate the ore deposits as a step towards mining. The AEC sponsored an intensive research program in collaboration with the U.S. Geological Survey to locate uranium deposits early in the Cold War (McKelvey 1955), and Helen Cannon led the bioprospecting research of this program on the Colorado Plateau in the U.S. where deposits were concentrated (Cannon 1952, 1957). Cannon found many species of higher plants that could indicate ore deposits and some have been found to be useful in restoring uranium mine tailings (May 1975) so there are segments of biodiversity adapted to low-level radioactivity. However, the artificial radiation created by humans with nuclear bombs and reactor accidents (Fetter and Tsipis 1981) go far beyond the background radiation to which species might have evolved adaptations.

Research in Cold War–era Radioecology illustrated how ecosystems self-organize in response to a novel stressor (radiation) for which they have little or no previous adaptation. Self-organization is the process by which various parts of a system connect themselves to work together when the system changes (Odum 1988, Levin 2005), such as with the addition of artificial radiation. In ecosystems self-organization relates to the composition and relative abundances of species in the system. The examples of ecological self-organization that emerged during the radiation experiments include both system collapse when species diversity is being reduced during radiation stress and system recovery when species diversity is increasing after the radiation stress was removed. During collapse, the species that survive radiation are termed radioresistant and during recovery, the species that recolonize a site that has been irradiated are termed radioresilient.

The records of the past studies from Cold War–era Radioecology are a resource for understanding the ecological consequences of pollution from radioactivity in today's world. For example, the "rewilding" that has occurred after reactor accidents at Chernobyl, in Ukraine, and at Fukushima, in Japan, could have been anticipated with knowledge of the AEC-funded ecology from the early Cold War years. In this book some effort will be given to comparing Radioecology from the Cold War–era period with the recent discipline.

Since the 1920s it has been known that artificial radiation could create mutations in the genome of an organism (Muller 1927, 1958). After the bombing of Japan at the end of World War II and during the early years of bomb testing, fears of radiation exposures in the general public generated myths in popular culture of mutant monsters created by radioactive pollution (Table 1.1). In essence this was the public's view of how ecosystems would respond to artificial radiation. Typically the monsters were destructive and killed humans, which was a reflection of the fears of radiation. These monsters were the product of human imagination, like the monsters thought to inhabit the New World that are often shown on European explorers' maps from the 1500s (Dickenson 1998). The fears of radioactivity are real for humans, who are much more sensitive to its toxicity than are whole ecosystems. However, in the history of Radioecology, when ecosystems were intentionally exposed to radiation, physical mutations were found to be a minor occurrence and no Godzillas ever appeared! This is because the selection pressure of mortality due to radiation stress (either direct or indirect) removes most organisms with mutations and favors resistant and resilient species through the selection processes of self-organization that are inherent in the ecosystem level of organization. Figure 1.2 depicts the two views of radiation effects with overview models in the energy circuit language (Odum 1994). The model diagram at the top of the figure shows the stress of radiation simplifying the ecosystem through mortality along with the imaginary view of radiation creating monsters that attack humans. Alternatively, the model diagram at the bottom of the figure shows the simplifying effect of mortality along with the action of radiation as a force that can select for species that are resistant

TABLE 1.1 Biodiversity of popular-culture movie monsters created by radiation during the early years of atomic bomb testing.

Year	*Name of the Movie*	*Monster*
1954	*Godzilla*	enormous, fire-breathing sea monster awakened by nuclear testing
1954	*Them!*	giant mutant ants spawned by nuclear testing
1955	*It Came From Beneath the Sea*	giant octopus awakened by nuclear testing
1957	*The Attack of the Crab Monsters*	radiation-mutated giant intelligent crabs

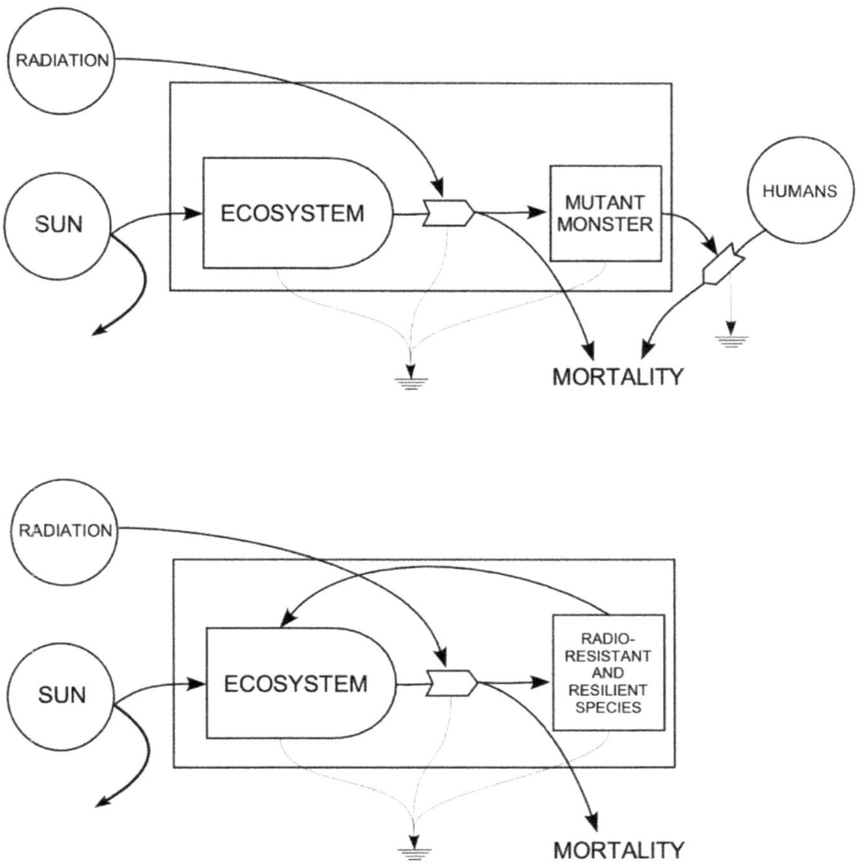

FIGURE 1.2 Minimodels of artificial radiation effects on ecosystems, contrasting the imaginary view of the mutant monster versus the more realistic view of radiation selecting for species that can resist or respond quickly after the stress is removed.

and resilient to the impact through self-organization. One of the goals of the book is to review how self-organization occurred in the various AEC-funded Radioecology projects and how this past research can relate to the understanding of changes that are creating novel ecosystems under present-day conditions of stress and disturbance.

As a final note, this book is a work of history, so the units reported are the traditional units used in the original publications. The rad is the traditional unit of absorbed dose of radiation and the gray (Gy) is the current unit, with 1 rad = 0.01 Gy. The rem is the traditional unit of biological dose of radiation or the dose equivalent and the current unit is the Sievert (Sv), with 1 rem = 0.01 Sv. The curie (Ci) is the traditional unit of radioactivity and the current unit is the Becquerel (Bq), with 1 Ci = 3.7 x 10^{10} Bq.

Literature Cited

Ayres, P. 2012. Shaping Ecology, The Life of Arthur Tansley. Wiley-Blackwell Publ., Oxford, UK.

Cannon, H. L. 1952. The effect of uranium-vanadium deposits on the vegetation of the Colorado Plateau. American Journal of Science 250:735–770.

Cannon, H. L. 1957. Description of indicator plants and methods of botanical prospecting for Uranium deposits on the Colorado Plateau. U.S. Department of the Interior, Geological Survey Bulletin 1030-M.

Cook, R. E. 1977. Raymond Lindeman and the trophic-dynamic concept in Ecology. Science 198:22–26.

Dickenson, V. 1998. Drawn From Life, Science and Art in the Portrayal of the New World. University of Toronto Press, Toronto, Canada.

Eisenbud, M. 1963. Environmental Radioactivity. McGraw-Hill Book Company, New York, NY.

Fetter, S. A. and K. Tsipis. 1981. Catastrophic releases of radioactivity. Scientific American 244(4):41–47.

Freedman, L. 2001. The Cold War. Cassell & Co., London, UK.

Klement, A. W., Jr. 1965. Radioactive fallout phenomena and mechanisms. Health Physics 11:1265–1274.

Kwa, C. 1993. Radiation ecology, systems ecology and the managing of the environment. Pp. 213–249. In: M. Shortland (ed.). Science and Nature. British Society for the History of Science, Oxford, UK.

Levin, S. A. 2005. Self-organization and the emergence of complexity in ecological systems. Bioscience 55:1075–1079.

Lindeman, R. L. 1942. The trophic-dynamic aspect of ecology. Ecology 23:399–418.

May, M. 1975. Reclamation of uranium-mined areas in the United States. Pp. 223–230. In: M. J. Chadwick and G. T. Goodman (eds.). The Ecology of Resource Degradation and Renewal. Blackwell Scientific Publications, Oxford, UK.

McKelvey, V. E. 1955. Search for Uranium in the United States. U.S. Department of the Interior, Geological Survey Bulletin 1030-A.

McMahon, R. J. 2003. The Cold War, A Very Short Introduction. Oxford University Press, Oxford, UK.

Muller, H. J. 1927. Artificial transmutation of the gene. Science 66:84–87.

Muller, H. J. 1958. General survey of mutational effects of radiation. Pp. 145–177. In: W. D. Claus (ed.). Radiation Biology and Medicine, Selected Reviews in the Life Sciences. Addison-Wesley Publ., Reading, MA.

Odum, H. T. 1988. Self organization, transformity, and information. Science 242:1132–1139.

Odum, H. T. 1994. Ecological and General Systems, An Introduction to Systems Ecology. University Press of Colorado, Niwot, CO.

OTA (Office of Technology Assessment). 1979. The Effects of Nuclear War. OTA-NS-89, Technology Assessment Board, Congress of the United States, Washington, DC.

Reichle, D. E. and S. I. Auerbach. 2003. U.S. Radioecology Research Programs of the Atomic Energy Commission in the 1950s. ORNL/TM-2003/280, Oak Ridge, TN.

Rhodes, R. 1986. The Making of the Atomic Bomb. Simon and Schuster Publ., New York, NY.

Rothschild, R. 2013. Environmental awareness in the Atomic Age: Radioecologists and nuclear technology. Historical Studies in the Natural Sciences 43:492–530.

Schull, W. J. 1995. Effects of Atomic Radiation, A Half-Century of Studies from Hiroshima and Nagasaki. Wiley-Liss Publ., New York, NY.

Sprugel, G., Jr. 1975. John N. Wolfe, 1910–1974. Bulletin of the Ecological Society of America 56:20–22.

Sterner, R. W. 2012. Raymond Laurel Lindeman and the Trophic Dynamic Viewpoint. American Society of Limnology and Oceanography, Waco, TX.

Tansley, A. G. 1935. The use and abuse of vegetational concepts and terms. Ecology 16:284–307.

Turner, M. G., V. H. Dale and E. H. Everham, III. 1997. Fires, hurricanes, and volcanoes: Comparing large disturbances. Bioscience 47:758–768.

Van der Valk, A. G. 2014. From formation to ecosystem: Tansley's response to Clements' climax. Journal of the History of Biology 47:293–321.

2

EARLY STUDIES IN RADIOECOLOGY

Introduction

There is an inherent irony in atomic (nuclear) energy in that it can have both positive and negative uses for humans. This was noted by Creager (2013) with the metaphorical statement that "the atom could cure as well as kill". This dichotomy played out in the early 1940s before the AEC was created and it continues today. In fact, atomic energy development began as a pursuit for positive uses of radiation under controlled conditions and only later, once World War II began, did it become a source of weapons of mass destruction. The earliest uses, that took place before the war, were in medicine and science. Once the bomb had been created and used to end World War II, governments strived to promote the positive or "peaceful" uses on both practical and political grounds, even as they actively participated in arms races to make bigger and better bombs. The environmental implications of some of the peaceful uses of nuclear energy will be covered later in the book, but in this chapter the focus is on Radioecology developments that took place before the AEC's vision of ecological studies began. In a sense these were precursors to the large-scale ecosystem studies funded by the AEC that are the main subject of this book.

Many of the uses of atomic energy (positive or negative) revolve around the production of radioisotopes that release energy (e.g., radiation). These are elements produced either in a machine (a cyclotron or nuclear reactor) under controlled conditions or in a bomb which releases the elements upon detonation. All atoms of the same element have the same atomic number, in terms of the number of protons in the nucleus. Isotopes are atoms of an element that have different numbers of neutrons and therefore different atomic mass, in terms of the number of protons plus neutrons in the nucleus. Unstable forms of isotopes are known as radioactive isotopes (also radioisotopes or radionuclides) and these spontaneously

DOI: 10.4324/9781003284901-3

undergo nuclear decay to a more stable state by emitting one or more forms of radiation. The three main kinds of radiation emitted during decay are called alpha particles, beta particles and gamma radiation. All radioactive isotopes emit nuclear radiation at a fixed rate measured in terms of half-life, which is the time necessary for half of the atoms in a sample of the element emit their radiation and become stable. Measurement of the radioactivity released during nuclear decay allows the tracing of radioisotopes or radionuclides in whatever their intended (or unintended) use. Therefore, Radioecology had its origins in tracing these elements through ecosystems even before the AEC was created.

Radioactive Fallout in Agricultural Food Chains

Tracking the fallout of radionuclides produced in bomb testing led to one of the earliest precursors of Radioecology and, in fact, it set the stage for much of Cold War–era Radioecology. Radioactive fallout was an unexpected development, as noted by Wargo (2009):

> Nuclear weapons were obviously intended to produce catastrophic environmental damage, but the Atomic Energy Commission's discovery of global fallout and ubiquitous human exposure was a surprise.

Although the AEC mandated safety regulations during bomb testing, the off-site spread of radioactive fallout became a problem that could not be ignored. Before the first bomb tests, it was assumed that radioactive debris from the detonation would fall out of the atmosphere to the ground surface quickly and nearby the blast site. This assumption proved to be correct for larger particles but it was wrong for smaller radioactive particles, which became suspended, and dispersed far beyond the blast site before falling out of the atmosphere. Fallout particles are injected into the lower atmosphere or troposphere by small nuclear weapons and typically come down to Earth within a few hundred miles or kilometers. Particles from large, thermonuclear weapons enter the upper atmosphere or stratosphere and circulate worldwide before falling down to the ground. Over time it was found that global circulation patterns kept most of the suspended particles in the Northern Hemisphere and that actual fallout patterns were complex.

The AEC became aware of the significance of radioactive fallout over the first few years of weapons testing but did not initially release its findings to the general public for political reasons (Divine 1978). According to Caufield (1990),

> In the early years of the test programme, fallout was largely a mystery. In some respects, it remained mysterious to the AEC longer than it did to others. The AEC's efforts to improve its scientific understanding of fallout were hampered by its military-political obligations. The commission's obsession with secrecy stifled a free exchange of information and ideas.

However, especially after the disastrous accident with the Bravo test of a thermonuclear bomb in March 1954, some journalists and scientists began to piece together the extent of this new form of pollution (see, for example, Lapp 1955). Consequently, the AEC, which had a monitoring program to measure the changes in radioactivity levels caused by nuclear testing, began releasing early reports on fallout (Eisenbud and Harley 1953, Andrews 1955).

Radioactive fallout was immediately feared to be a threat to human health, especially through contamination of foods. This fear brought attention to an important ecological concept: the food chain. Of particular interest were food chains that have humans as the last link. At this point, in the early to mid-1950s, knowledge of food chains in natural ecosystems was minimal, so the academic discipline of Ecology had little to contribute to the hazard assessment of radioactive fallout to humans. Instead, agronomists and nuclear scientists began examining agricultural food chains with an interest in potential movement of fallout radionuclides up the food chain to humans. Figure 2.1 is an early conceptual systems diagram illustrating the focus of radionuclides as a threat to humans. In a sense the black boxes in the diagram connote what at the time was the uncertainty in the state-of-the-art understanding of the movements of radionuclides through the ecosystem.

Although many radionuclides are found in fallout from an atomic bomb detonation, initial interest was on three most likely to harm humans. Strontium-90 is considered to be the most hazardous fission product because of its relatively long half-life (28 years) and because it chemically resembles calcium. Thus, as noted by Caster (1960), "when strontium-90 falls to earth, it becomes mixed with calcium and follows calcium through its ecological chain in Nature". Sr-90 reaches humans through milk and vegetables and becomes incorporated into bone (because of its similarity to calcium) where it remains as a source of internal radiation in the body. Cesium-137, also a fission product, chemically resembles potassium and it becomes distributed throughout the body. It has a physical half-life similar to Sr-90 at 30 years but its biological half-life is short, at about 17 days, since it gets excreted quickly through normal metabolism rather than being retained in the body, making it is less of a threat for internal radiation than Sr-90. Another fission product, Iodine-131, has a half-life of only eight days but it is produced in relatively large amounts in a nuclear detonation. It moves through the food chains and concentrates in the thyroid gland of mammals, where it becomes a source of internal radiation. All of these radionuclides can cause cancer in animals, so they represent serious threats to human health from radioactive fallout.

Extensive studies of agricultural food chains were undertaken, which showed how fallout radionuclides were taken up from the soil by plants or how they became absorbed onto leaf surfaces. The radionuclides could be directly eaten when plants are consumed in the human diet or the contaminated plants could be eaten by grazing animals, which are then eaten as part of the human diet. A special focus was given to the forage-to-cow-to-milk pathway. In some sense the fear that milk

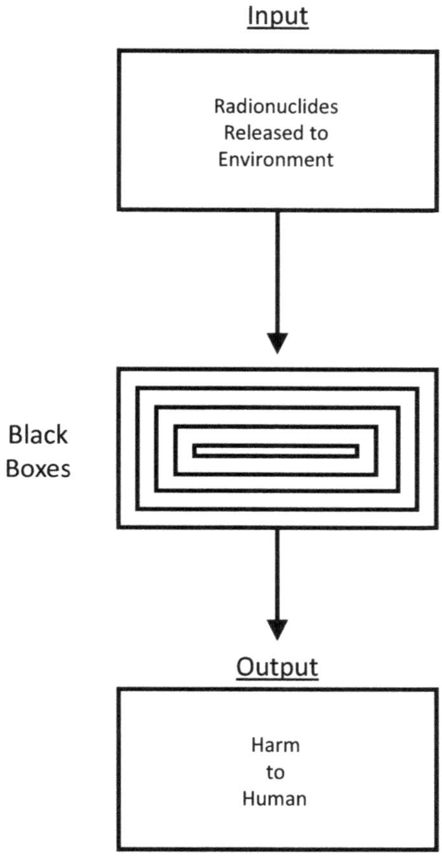

FIGURE 2.1 Overview model of the movement of radionuclides through the environment to humans (from Comar and Lengemann 1966). The environment is shown as a series of nested black boxes that represent the ecosystem.

Source: used with permission from elsevier.

could be radioactive seems to have been particularly frightening and of special concern to the general public. Figure 2.2 shows the details of food chains in an agricultural system in which fallout radionuclides enter from the atmosphere and move along the pathways to humans. Much research was undertaken to quantitatively assess this general model. Leaders in this work were C. L. Comar at the Oak Ridge National Laboratory and R. S. Russell at Oxford University, in England, and early work was published in symposium proceedings (Fowler 1965, Caldecott and Snyder 1960, Russell 1966, Åberg and Hungate 1967) and elsewhere. Because of the importance of understanding the dynamics of fallout radionuclides in agricultural food chains, the work continued throughout and after the Cold War years and knowledge has become sophisticated (see, for example, Desmet 1988, Whicker and Kirchner 1987).

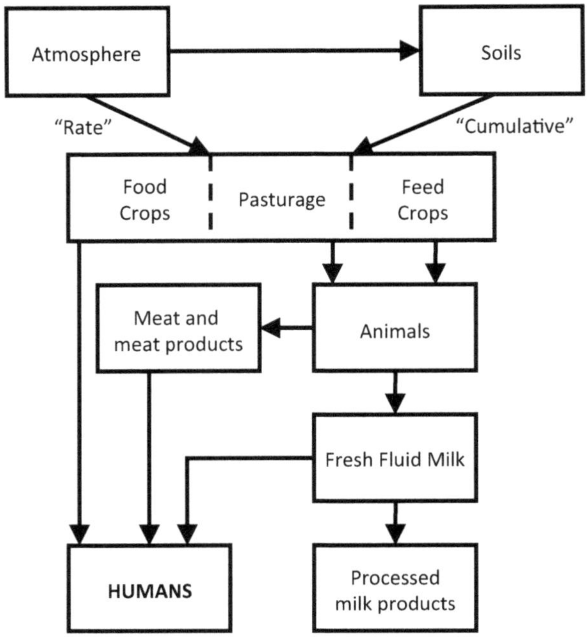

FIGURE 2.2 A compartment model of an agricultural food web with humans as the "top predator" in the system (from Comar and Lengemann 1966).

Source: Used with permission from Elsevier.

Ecologists began studying radionuclides in natural ecosystems after the initial work done in agriculture. They benefited from the early work on radionuclides in agricultural systems by the development of measurement techniques and the understanding of how radionuclides enter and move through food chains. Examples of early studies of the movements of radionuclides through natural ecosystems by ecologists were reported in the first *Proceedings of the United Nations International Conference on the Peaceful Uses of Atomic Energy* (Foster and Davis 1956, Hansen and Kornberg 1956).

Significant human impacts did arise in localized areas with heavy exposure to fallout, such as in the rural communities immediately downwind from the Nevada Test Site (Fradkin 1989, Fuller 1984) and in the Marshall Islands from the Bravo test in 1954 (see Chapter 3). However, though of great concern to humans, radioactive fallout was thought to be "never high enough to noticeably affect ecosystem structure or function" (Jordan 1985), which is the focus of this book. Ultimately, most nuclear powers agreed to eliminate atmospheric testing of weapons, starting in 1963, which greatly reduced radioactive fallout.

Tracing Radioactive Phosphorus in Lakes

Scientists and physicians began using artificially produced radiation sources as soon as they were available from the operation of cyclotrons before World War

II. Radioisotopes were used as molecular tracers in both research and medicine. These isotopes allowed compounds to be intentionally tagged or labeled by replacement of an ordinary, stable atom with its corresponding radioactive isotope. The compound could then be traced through chemical reactions or biological systems by detecting the radiation emitted from the radioisotope as it decayed. Creager (2013) has provided an extensive review of the positive use of radioisotopes in many applications. The availability of radioisotopes was limited before and during World War II, but their production greatly expanded after the war when the nuclear reactors built to produce bomb materials could also produce radioisotopes for other uses, such as medical applications. For example, Creager (2013) notes, "From 1946 to 1955, the AEC's Oak Ridge contractor sent out nearly 64,000 shipments of radioactive materials to more than 2,400 laboratories, companies, and hospitals." This is a prime example of the "peaceful uses" of atomic energy that became part of the AEC's commercial and political agenda once it became established. Thus, Creager (2013) aptly describes the radioisotopes used by researchers and physicians as "by-products" of the bomb and, by extension, of the Cold War.

The first use of radiotracers in Ecology involved tracing the phosphorus cycle in freshwater lakes in the 1940s and 1950s. Phosphorus-32 could be produced in nuclear reactors and it was particularly attractive as a research tool because of its short half-life of 14 days. It could be safely introduced into Nature because the radioactivity would be quickly decayed, as long as the amounts added were small. Phosphorus is an important element in biological systems, so P-32 was quickly embraced by Ecologists as a radiotracer. Phosphorus is a limiting factor for primary productivity in most freshwater lakes, so it was understandable that limnologists (ecologists who study lake ecosystems) were the first to use P-32 as a research tool.

The first ecologist to intentionally add a radioactive tracer into an ecosystem was G. E. Hutchinson of Yale University, in Connecticut. Hutchinson was probably the most famous ecologist of the twentieth century, and his contributions are well known (see, for example, Deevey 1972, Slack 2010, Skelly et al. 2010). He was a true scholar and a prolific writer (for example, he published four volumes of essays along with his scientific publications) with broad interests, but he is best known as a limnologist—as an example, his *Treatise on Limnology* was 3,734 pages long, published in four volumes. Starting in the late 1930s he and his students carried out detailed studies of the biology, paleohistory and biogeochemistry of a small lake near the Yale campus, named Linsley Pond. His motivation was to understand the structure and function of the lake as a system in multiple dimensions.

Linsley Pond, like many lakes in the North Temperate Zone, becomes thermally stratified in the summer, due to density differences caused by temperature differences. Under these conditions, the waters in the upper and warmer, lighted layer (termed the epilimnion) do not mix with the waters in the lower and colder layer (termed the hypolimnion) until temperatures drop in the Fall and the water column mixes until the following Spring. When the lake is stratified in the Summer, phosphorus concentrations in the epilimnion decline as the nutrient is taken up by plankton, while in the hypolimnion phosphorus concentrations increase due to sedimentation from the epilimnion and exchanges between the water and the bottom muds. In the Fall the phosphorus concentration in the epilimnion is restored when the stratification breaks down and the lake "turns over", mixing the surface and bottom waters. Thus, phosphorus was thought to cycle primarily on a seasonal basis with alternating mixing and stratification.

In the late 1930s Hutchinson understood that both physical and biological processes were taking place in the lake but the cycling of nutrients had not yet been articulated. The phosphorus cycle was clearly critical to the overall functioning of the lake as an ecosystem, since phosphorus was a nutrient in short supply that limited growth of phytoplankton, but the available techniques for studying the cycle only allowed the concentrations in various compartments to be measured at different times of the year. Under these circumstances, only a crude understanding of the phosphorus cycle was possible. In a important paper in 1941 Hutchinson reported a series of integrated studies in the lake pointing at the overall "metabolism" of the system using phosphorus cycling as an index (Hutchinson 1941). Based on measurements of water movements and chemistry he inferred the pattern of the phosphorus cycles and speculated on uptake rates by the phytoplankton in the lighted epilimnion and higher plants in the littoral zone of the lake, regeneration rates from bacteria in lake muds of the hypolimnion along with multiple horizontal and vertical physical flows of phosphorus. From his measurements and reasoning about the lake as a closed system, he noted that "variations in the total phosphorus content of the whole lake are considerable and rapid" with estimated weekly fluxes of about a kilogram added (through regeneration) and lost (through uptake and sedimentation) from the lake waters. This was an important paper, but it was probably a bit unsatisfying for Hutchinson since it was based on a number of assumptions and inferences. The availability of P-32 allowed him to advance the study of the phosphorus cycle dramatically and he quickly adopted the radioisotope tracer method as soon as was possible for him. He first acquired a small amount of P-32 from E. F. Pollard, a physicist who operated a cyclotron at Yale, and added it to Linsley Pond in the summer of 1941, but the quantity was too small to trace with available measurement techniques. After World War II he followed up on this initial study in 1946, again with P-32 from Pollard, and also in 1947, this time with radionuclides provided by the AEC from its nuclear reactors. In all of these studies the phosphorus isotope was added to waters from a rowboat in a transect across the lake. Measurements of P-32 in various lake compartments were made weekly, but the experiments could only last for about a month due to the rapid decay rate of the radioisotope. Through this work he verified the movements of phosphorus in the lake and was better able to demonstrate the fast rates of movements that he had earlier suspected (Hutchinson and Bowen 1947, 1950).

The significance of Hutchinson's work with P-32 was described in a volume of *Benchmark Papers in Ecology* by Pomeroy (1974):

> The labelling of Linsley Pond had two effects on ecology. Its general effect was to set in motion a series of experiments by other investigators on whole

> lakes. These were followed by experiments on whole watersheds and whole landscapes. Investigators came to realize that they could treat whole ecosystems as functioning units, just as the physiologist deals with whole organisms, and that they could do experiments at the ecosystem level. This attitude existed already in biological oceanography, where the massive size of the system made detailed approaches less appealing, but Hutchinson and Bowen brought the whole-system experiment out of the sea and pointed it toward the land.

Extensive studies of phosphorus cycling in lakes with P-32 were independently undertaken by Canadian research groups starting in the late 1940s and early 1950s. A group at Dalhousie University, in Halifax, Nova Scotia, started experiments in 1949 and a group at the University of Toronto, in Ontario, started in 1951. Both of these groups found rapid movements of the added radioisotope tracer through the compartments of the lake ecosystem and showed the importance of bacteria in the phosphorus cycle. The motivations for the research included an interest in understanding how fish production in the lakes could be increased by fertilization, since phosphorus was known to be a limiting factor for aquatic primary productivity (Smith 1966). Much of this research was supported by Atomic Energy of Canada Limited (which was similar to the AEC in the United States) along with the National Research Council of Canada. Research was carried out using P-32 as a radiotracer in different kinds of lakes and with a number of ingenious microcosm designs that mimicked the muds in the lake bottoms, which were known to be important in phosphorus cycling. Figure 2.3 is a systems diagram of the phosphorus movements through the compartments of the lake, showing the rapid cycling rates that were unsuspected before the availability of P-32.

In summary, the early work with P-32 in freshwater lakes was an important precursor to the ecosystem-based research funded by the AEC. The P-32 radioisotope became widely used by ecologists as a tracer, but it was especially useful to limnologists who needed to understand the cycling in lakes where phosphorus is a limiting nutrient. G. E. Hutchinson began using the isotope even before the AEC came into existence. Table 2.1 shows the early, essentially courageous, experiments where artificial radioactivity was intentionally added to natural ecosystems for ecological studies. The limnologists basically had to invent experimental designs and techniques to incorporate P-32 into their research, but it was worth the effort. As an example of their inventiveness, since most of the early studies added the tracer to the surface water, the Dalhousie University group created a special bottle which contained a dynamite cap to add the isotope in the bottom waters (McCarter et al. 1952). In operation, the bottle was lowered into the bottom waters where it was exploded electronically through wires attached to a battery in the boat on the lake surface, thereby releasing the P-32 solution into the hypolimnion!

It was impossible to answer the kinds of questions that the limnologists were asking about lakes as systems until P-32 became available, so the radiotracer

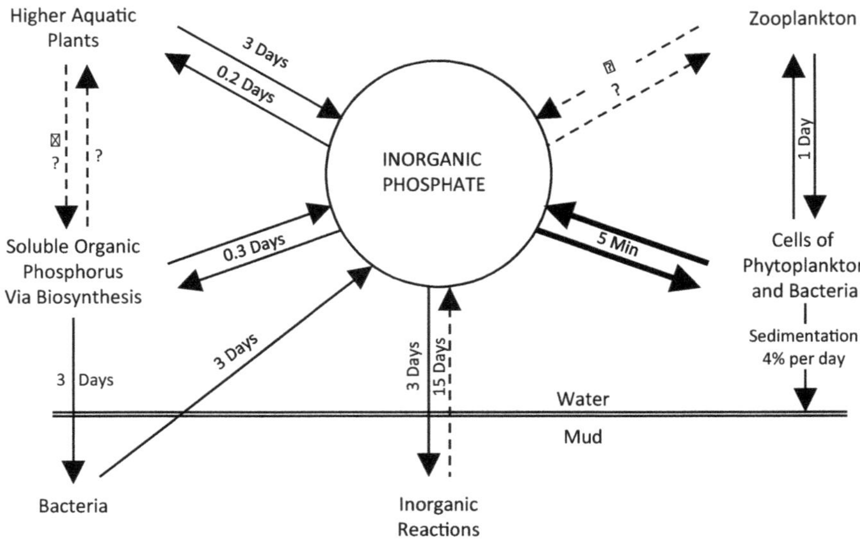

FIGURE 2.3 An early systems diagram of the phosphate cycle in a freshwater lake, showing the speed of movements of the P-32 radioisotope along the pathways (from Hayes and Phillips 1958).

Source: Used with permission from John Wiley & Sons.

TABLE 2.1 Comparison of early work where P-32 was added to freshwater lakes in order to study the phosphorus cycle.

Lake Name	*Area (ha)/ Depth (m)*	*Date*	*Amount of P-32 Added*	*Reference*
Linsley Pond, Connecticut	9.4 ha/14.8 m	1946	10 millicuries	Hutchinson and Bowen (1947)
Linsley Pond, Connecticut	9.4 ha/14.8 m	1947	70 millicuries	Hutchinson and Bowen (1950)
Punchbowl, Nova Scotia	0.3 ha/6.2 m	1948	100 millicuries	Coffin et al. (1949)
Toussaint, Ontario	4.7 ha/9.8 m	1951	200 millicuries	Rigler (1956)

technique was invaluable for ecological research. The speed at which phosphorus moved and the pathways it moved along were unexpected and these findings demonstrated a new view of the dynamism of these ecosystems.

Hutchinson did no further work with radionuclides beyond his research with P-32 at Linsley Pond in the 1940s. However, the Canadian groups built on their early work and studied many more lakes and extended the work with P-32 to marine waters. Other limnologists, including A. D. Hasler at the University of

Wisconsin, also worked with P-32 and other radionuclides with funding from the AEC once it was established. For example, Hasler's doctoral student G. E. Likens used radiosodium to study water movements and cycling in lakes (Likens and Hasler 1960, 1962). Likens went on to initiate the important Hubbard Brook watershed project (Holmes and Likens 2016), which became a model for ecosystem research (Likens 1992). He represents another example of a leader in Ecology whose early training was in Radioecology with support from the AEC.

Carbon-14 and the Primary Production of the Oceans

Another precursor to Radioecology's development involved the use of the Carbon-14 radioisotope as a tracer. Carbon is the most critical element to living systems, so the availability of a tracer isotope for it was a major research breakthrough. C-14 is produced naturally in the upper atmosphere by the bombardment of nitrogen by neutrons from cosmic rays, but it makes up a tiny percentage (about one part in a million million) of the total carbon on Earth. This radioisotope can also be made by a similar process in a cyclotron or nuclear reactor (or in an atomic bomb detonation), so it became available to scientists, especially after World War II. It was proven to exist in Nature by a research group led by Willard F. Libby that developed a special Geiger counter for its measurement. In 1949 Libby created the radiocarbon dating method, for which he won the Nobel Prize in 1960, by comparing the ratio of carbon isotopes (C-14 relative to C-12) in ancient specimens of organic matter (Libby 1952). This method takes advantage of the fact that all organisms take up the carbon isotopes in an equilibrium (or constant ratio) while living, but upon death the ratio begins to change as the amount of C-14 disintegrates through radioactive decay. Fortunately, the decay rate is slow, since the half-life of C-14 is 5,730 years. Radiocarbon dating can be used on organic specimens as old as about 55,000 years. Beyond that point the amount of C-14 remaining is too low for accurate measurement. This method has been an invaluable tool for dating in archeology and historical ecology.

As an aside, Libby was an interesting personality in the history of the atomic bomb. He was a veteran of the Manhattan Project and, after his innovative work with C-14, he was appointed to the General Advisory Committee of the AEC. In 1954 he became one of the actual commissioners. Nicknamed "Wild Bill", he was politically active and a vocal supporter of the development of nuclear weapons in the Cold War. Libby became in the late 1950s the "leading spokesman for the view that fallout posed minor hazards at worst" (Hacker 1994), a position which ultimately turned out to be incorrect.

Once C-14 was produced artificially, it was quickly used in studies of photosynthesis (Creager 2013). Before World War II, C-14 produced in cyclotrons was used to track uptake and movements of carbon in the photosynthesis process. The main work was done by Melvin Calvin and his associates (Bassham and Calvin 1957) using laboratory cultures of the microalgae *Chlorella* as a test organism. They discovered the steps plants use to make sugar out of carbon dioxide that had been labeled with C-14 and they received the Nobel Prize for the work in 1961.

Einer Steemann Nielsen was the first to use the C-14 tracer in ecosystems to measure primary productivity (Steemann Nielsen 1952). He acquired the isotope from the Oak Ridge National Laboratory and used it for measuring the rate of primary production in the sea during the Danish Galathea expedition around the world during the years 1950–1952 (Bruun et al. 1956). The approach was essentially to take into the field (in this case the open ocean) the method used by Calvin and others for studying photosynthesis in the laboratory. The C-14 method is now well known and widely used in aquatic ecology (see, for example, Vollenweider 1974, Sorokin 1999). In outline, the C-14 isotope is added as biocarbonate to a sample bottle of seawater, which is then either suspended in the ocean or placed in a lighted incubator on board the ship. After a certain time interval (typically four hours) the plankton organisms in the water of the sample bottle are removed by filtration and the amount of C-14 that was taken up is measured by the beta radiation emitted by the organisms retained on the filter. Radiation measurements were first done with a Geiger-Mueller counter, but more recently a liquid scintillation counter is used. With suitable corrections the carbon uptake is calculated based on the beta emissions.

Steemann Nielsen estimated productivity at 184 stations during the round-the-world cruise over the photic zone of the open ocean that ranged up to 120 meters (394 ft) deep (Steemann Nielsen 1955). He classified the ocean waters into four regions based on water mixing and fertility, with productivities ranging from a high of about 3 g C/m2/day in areas that are nutrient rich to a low of about 0.05 g C/m2/day for nutrient-poor areas. Based on the results from the cruise he estimated the net production of the world's oceans to be about 1.5 x 10^{16} tons of carbon per year, which was similar to production estimates for the land. He continued to work on the method for decades (Steemann Nielsen 1975) with the intention of improving the estimates of primary productivity.

The C-14 method for measuring primary productivity in both saltwater and freshwater ecosystems was widely adopted by ecologists because of its sensitivity and because it was simple and fast, relative to other available methods. Many problems with the method were quickly identified as more researchers applied it, and its limitations have been qualified over time (see, for example, Peterson 1980, Falkowski and Raven 2007). However, the quote by Morris et al. (1971) summarizes what many researchers feel about the method:

> When Steemann Nielsen (1952) introduced the ^{14}C technique, many of us actively engaged in trying to assess the level of primary production breathed a sigh of relief. This was a method that we had waited for, one that could provide fundamental information about the incorporation of carbon into aquatic ecosystems. Despite many minor ups and downs in

general methodology, this technique has survived and has been widely accepted by its practitioners.

Finally, as an aside, it is interesting to note that C-14 went to Mars in 1975 as part of NASA's Viking Lander program. The radioisotope was used in one of the robotic tests for the existence of life in Martian soils. C-14 was injected into a small enclosed chamber on the lander that contained Martian soil and checked with an onboard Geiger counter for uptake that would suggest biological metabolism. Uptake did take place, but it was later judged to have occurred due to inorganic chemical reactions rather than extraterrestrial metabolism (Horowitz et al. 1977). While this was an obviously disappointing result, the Viking Lander program was a major victory for the U.S., which was in a space race with the Soviet Union as part of the Cold War (Cadbury 2006).

Literature Cited

Åberg, B. and F. P. Hungate (eds.). 1967. Radioecological Concentration Processes. Pergamon Press, Oxford, UK.

Andrews, H. L. 1955. Radioactive fallout from bomb clouds. Science 122:453–456.

Bassham, J. A. and M. Calvin. 1957. The Path of Carbon in Photosynthesis. Prentice-Hall, Inc., Englewood Cliffs, NJ.

Bruun, A. F., Sv. Greve, H. Mielche and R. Sparck (eds.). 1956. The Galathea Deep Sea Expedition 1950–1952. George Allen and Unwin LTD, London, UK.

Cadbury, D. 2006. Space Race, The Epic Battle Between America and the Soviet Union for Dominion of Space. HarperCollins Publishers, New York, NY.

Caldecott, R. S. and L. A. Snyder. 1960. Radioisotopes in the Biosphere. Center for Continuation Study, University of Minnesota, Minneapolis, MN.

Caster, W. O. 1960. From bomb to Man. Pp. 37–50. In: J. M. Fowler (ed.). Fallout, A Study of Superbombs, Strontium 90, and Survival. Basic Books, Inc., New York, NY.

Caufield, C. 1990. Multiple Exposures, Chronicles of the Radiation Age. University of Chicago Press, Chicago, IL.

Coffin, C. C., F. R. Hayes, L. H. Jodrey and W. G. Whiteway. 1949. Exchange of materials in a lake as studied by the addition of radioactive phosphorus. Canadian Journal of Research 27d:207–222.

Comar, C. L. and F. W. Lengemann. 1966. General principles of distribution and movement of artificial fallout through the biosphere to man. Pp. 1–18. In: B. Åberg and F. P. Hungate (eds.). Radioecological Concentration Processes. Pergamon Press, Oxford, UK.

Creager, A. N. H. 2013. Life Atomic, A History of Radioisotopes in Science and Medicine. University of Chicago Press, Chicago, IL.

Deevey, E. S. (ed.). 1972. Growth by Intussusception, Ecological Essays in Honor of G. Evelyn Hutchinson. Transactions of the Connecticut Academy of Arts and Sciences 44:1–443.

Desmet, G. (ed.). 1988. Reliability of Radioactive Transfer Models. Elsevier Applied Science, London, UK.

Divine, R. A. 1978. Blowing on the Wind, The Nuclear Test Ban Debate 1954–1960. Oxford University Press, New York, NY.

Eisenbud, M. and J. H. Harley. 1953. Radioactive dust from nuclear detonations. Science 117:141–147.

Falkowski, P. G. and J. A. Raven. 2007. Aquatic Photosynthesis, Second Edition. Princeton University Press, Princeton, NJ.
Foster, R. F. and J. J. Davis. 1956. The accumulation of radioactive substances in aquatic forms. Pp. 364–367. In: Proceedings of the International Conference on the Peaceful Uses of Atomic Energy. Volume 13. United Nations, New York, NY.
Fowler, E. B. (ed.). 1965. Radioactive Fallout, Soils, Plants, Foods, Man. Elsevier Publishing Co., Amsterdam.
Fradkin, P. L. 1989. Fallout: An American Nuclear Tragedy. University of Arizona Press, Tucson, AZ.
Fuller, J. G. 1984. The Day We Bombed Utah: America's Most Lethal Secret. New American Library, New York, NY.
Hacker, B. D. 1994. Elements of Controversy: The Atomic Energy Commission and Radiation Safety in Nuclear Weapons Testing 1947–1974. University of California Press, Berkeley, CA.
Hansen, W. C. and H. A. Kornberg. 1956. Radioactivity in terrestrial animals near an atomic energy site. Pp. 385–388. In: Proceedings of the International Conference on the Peaceful Uses of Atomic Energy. Volume 13. United Nations, New York, NY.
Hayes, F. R. and J. E. Phillips. 1958. Lake water and sediment IV. Radiophosphorus equilibrium with mud, plants, and bacteria under oxidized and reduced conditions. Limnology and Oceanography 3:459–475.
Holmes, R. T. and G. E. Likens. 2016. Hubbard Brook: The Story of a Forest Ecosystem. Yale University Press, New Haven, CT.
Horowitz, N. H., G. L. Hobby and J. S. Hubbard. 1977. Viking on Mars: The carbon assimilation experiments. Journal of Geophysical Research 82:4659–4662.
Hutchinson, G. E. 1941. Limnological studies in Connecticut IV: The mechanisms of intermediary metabolism in stratified lakes. Ecological Monographs 11:21–60.
Hutchinson, G. E. and V. T. Bowen. 1947. A direct demonstration of the phosphorus cycle in a small lake. Proceedings of the National Academy of Sciences of the United States of America 33:148–153.
Hutchinson, G. E. and V. T. Bowen. 1950. Limnological studies in Connecticut. IX: A quantitative radiochemical study of the phosphorus cycle in Linsley Pond. Ecology 31:194–203.
Jordan, C. F. 1985. Nutrient Cycling in Tropical Forest Ecosystems. John Wiley & Sons, Chichester.
Lapp, R. E. 1955. Radioactive Fall-out. Bulletin of Atomic Scientists 11:45–51.
Libby, W. F. 1952. Radiocarbon Dating. University of Chicago Press, Chicago, IL.
Likens, G. E. 1992. The Ecosystem Approach: Its Use and Abuse. Ecology Institute, Oldendorf/Luhe, Germany.
Likens, G. E. and A. D. Hasler. 1960. Movements of radiosodium in a chemically stratified lake. Science 131:1676–1677.
Likens, G. E. and A. D. Hasler. 1962. Movements of radiosodium (Na24) within an ice-covered lake. Limnology and Oceanography 7:48–56.
McCarter, J. A., F. R. Hayes, L. H. Jordey and M. L. Cameron. 1952. Movement of materials in the hypolimnion of a lake as studied by the addition of radioactive phosphorus. Canadian Journal of Zoology 30:128–133.
Morris, I., C. M. Yentsch and C. S. Yentsch. 1971. Relationship between light carbon dioxide fixation and dark carbon dioxide fixation by marine algae. Limnology and Oceanography 16:854–858.

Peterson, B. J. 1980. Aquatic primary productivity and the $^{14}C-CO_2$ method: A history of the productivity problem. Annual Reviews of Ecology and Systematics 11:359–385.
Pomeroy, L. R. 1974. Cycles of Essential Elements. Dowden, Hutchinson & Ross, Inc., Stroudsburg, PA.
Rigler, F. H. 1956. A tracer study of the phosphorus cycle in lake water. Ecology 37:550–562.
Russell, R. S. (ed.). 1966. Radioactivity and Human Diet. Pergamon Press, Oxford, UK.
Skelly, D. K., D. M. Post and M. D. Smith (eds.). 2010. The Art of Ecology, Writings of G. Evelyn Hutchinson. Yale University Press, New Haven, CT.
Slack, N. G. 2010. G. Evelyn Hutchinson and the Invention of Modern Ecology. Yale University Press, New Haven, CT.
Smith, M. W. 1966. The Atlantic Provinces of Canada. Pp. 521–534. In: D. G. Frey (ed.). Limnology in North America. University of Wisconsin Press, Madison, WI.
Sorokin, Y. I. 1999. Radioisotopic Methods in Hydrobiology. Springer-Verlag, Berlin.
Steemann Nielsen, E. 1952. The use of radioactive carbon (C14) for measuring organic production in the sea. J. Cons. Int. Explor. Mer. 18:117–140.
Steemann Nielsen, E. 1955. Production of organic matter in the oceans. Journal of Marine Research 14:374–386.
Steemann Nielsen, E. 1975. Marine Photosynthesis. Elsevier Sci. Publ. Co., Amsterdam.
Vollenweider, R. A. 1974. A Manual on Methods for Measuring Primary Production in Aquatic Environments. IBP Handbook No. 12, Second Edition. Blackwell Scientific Publ., Oxford, UK.
Wargo, J. 2009. Green Intelligence, Creating Environments that Protect Human Health. Yale University Press, New Haven, CT.
Whicker, F. W. and T. B. Kirchner. 1987. PATHWAY: A dynamic food-chain model to predict radionuclide ingestion after fallout deposition. Health Physics 52:717–737.

PART II

3

THE PACIFIC PROVING GROUNDS

Introduction

The power of the atomic bombs used at the end of World War II convinced the United States government that these kinds of weapons were important for the future of warfare. Therefore, the industrial complexes developed during the Manhattan Project to produce atomic bombs were maintained and a plan for developing and testing future bombs was formulated. Testing was necessary because the technology was so new and, if the weapons were to be used, their performance needed to be predictable and thoroughly understood. The main criteria for choosing a site for nuclear weapons testing were that the location should be remote so that testing could be done with minimal risk and hazard to human life and it should be under American control. Many potential sites were considered, but ultimately the Marshall Islands, essentially in the middle of the Pacific Ocean, were chosen after World War II.

The Marshall Islands are a collection of atolls or coral reef–dominated islands located nearly 4,800 kilometers (3,000 miles) south of Hawaii. There was relatively little scientific information about Pacific atolls before the start of the atomic bomb testing, but they would be studied to help understand the effects of the testing program. Donaldson (1960) provided a general description of the atoll environment:

> An atoll may be described as a roughly oval, coralline reef rising 15,000 feet above the ocean floor. Within the surrounding reef there is enclosed a shallow lagoon generally with a maximum depth of about 180 feet. The lagoon is open to the ocean by one or more passes cutting through the reef, most of which is awash except at low tides. Emergent land consists of low sandy islands with an elevation seven to ten feet above sea level; elevations as high as twenty feet are rare. The islands occupy only a small fraction of

DOI: 10.4324/9781003284901-5

> the total area of an atoll. Bikini lagoon covers 229 square miles . . . and has a dry land area of two and a third square miles divided among some thirty-six islands and islets; Eniwetok lagoon covers 388 square miles, has a dry land area of two and one-fourth square miles, and forty islands.

Most of the bomb testing on the Pacific Proving Grounds took place on two of the Marshall Islands: Bikini and Eniwetok Atolls. However, several other atolls were affected by radioactive fallout from the tests. Several years after testing on the proving grounds had been completed, a few additional tests were done on Johnson Island, located about 160 kilometers (100 miles) south of Hawaii.

A number of the atolls of the Marshall Islands had been occupied by the Japanese during World War II and several battles took place there. After the war these islands, along with others in the Pacific Ocean, became a United Nations Trust Territory administered by the U.S. (Hezel 1995), which facilitated their choice as the location for the Pacific Proving Grounds.

It was decided that the first tests carried out at Bikini Atoll would evaluate the effects of atomic bombs on naval ships. The code name for the operation was Crossroads and it involved two bombs: the first dropped from a plane (test Able) and the second detonated underwater. These tests were carried out in July 1946, after the Marshallese residents were relocated to Rongerik Atoll, as a large-scale public demonstration for foreign diplomats, journalists, U.S. government personnel and scientists. The bombs were approximately the same size as those used at Hiroshima, Japan. A second set of tests, code-named Sandstone, was carried out without the public demonstration component in April and May 1948. These tests took place at Eniwetok Atoll, each detonated on towers for control of elevation, timing and measurement of the explosion.

After Operation Sandstone a short break in testing occurred. However, progress in the Cold War brought a rapid return to testing with Operation Greenhouse at Eniwetok in 1951. The most important development in the Cold War was the first test by the Soviets, in August 1949 in Kazakhstan. This test was named "First Lightning" by the Soviets (and "Joe 1", in reference to Joseph Stalin, by the Americans). This bomb was similar to the plutonium bomb dropped on Nagasaki, Japan, at the end of World War II. Additionally, developments in the delivery systems for the bomb were taking place rapidly as part of the arms race, with aircraft bombers being supplemented by ICBMs and nuclear submarines (McMahon 2003). These developments intensified the need to test more powerful bombs at the Pacific Proving Grounds.

Early in 1950 President Truman decided to test the thermonuclear hydrogen bomb, also known as the "super", and most of the subsequent experimentation involved this weapon. Operation Ivy on Eniwetok began in late 1952, which included the detonation of the first hydrogen bomb, code-named Mike. This explosion had the force of 10.4 megatons of TNT, 750 times greater than the Hiroshima bomb.

The largest detonation, code-named Bravo, occurred on Bikini in March 1954. This hydrogen bomb had the force of 15 megatons of TNT, which was larger than had been predicted beforehand. This oversight was compounded by an unanticipated shift in the winds during the test that caused fallout to spread to the nearby islands of Rongelap, Rongerik and Utrik, where the native Marshallese people (along with some U.S. servicemen stationed there) were contaminated by the radiation and forced to evacuate the islands. Additionally, a Japanese fishing boat, the *Fukuryu Maru* (*Lucky Dragon*), that was operating in water more than 200 kilometers (124 miles) from the detonation, was contaminated by fallout. By the time the fishing boat reached port in Japan, the crew were suffering from radiation poisoning, and ultimately, one of them died from the contamination (Matashichi 2011). Overall, the Bravo test turned into an international incident by revealing the dangers of radioactive fallout (Hines 1962). The U.S. received criticism from around the world and it seems the Bravo test was a social turning point that eventually led to a ban on atmospheric testing of atomic and thermonuclear bombs (Rothschild 2013).

The bomb testing on the Pacific Proving Grounds continued until 1958, when the work shifted to the Nevada Test Site (see Chapter 4). The U.S. government reported that 23 bombs had been detonated on Bikini and 43 on Eniwetok during the period from 1946 to 1958. The toll of these tests on the native Marshallese people, who were displaced and in some cases contaminated, is a terrible story that has been reported by multiple authors (Dibblin 1988, Kiste 1974), and it is a continuing issue of concern.

Monitoring at the Pacific Proving Grounds

Monitoring of the environmental effects of nuclear bomb detonations at the Pacific Proving Grounds took place unevenly throughout the testing period from 1946 to 1958. Personnel from the Applied Fisheries Laboratory (AFL) at the University of Washington were involved from the start and they more or less coordinated the overall monitoring efforts. Their work was described by Neal Hines, who was a science writer for the Atomic Energy Commission (see, for example, Hines 1966) and a participant in some of the monitoring expeditions. His main text (Hines 1962) gives a complete and detailed record of the testing program and the monitoring work. Additional details are given by Rothschild (2013) and Martin (2018), who surveyed unpublished reports of the AFL to the AEC and who interviewed some of the AFL personnel about their work.

The AFL was chosen to lead the environmental monitoring of the nuclear bomb testing because of its previous connection to studies of radiation effects on fish populations in the Columbia River from the Hanford complex during World War II. Lauren Donaldson formed the AFL and studied the effects of X-rays on salmon with funding from the Manhattan Project before the end of the war. Rothschild (2013) suggests General Leslie Groves, head of the Manhattan Project, had personally chosen the University of Washington fisheries group because of his concern for potential impacts on the Columbia River since he had grown up in the area and had attended the University of Washington. Because of this connection, in 1946 the AFL was included in Operation Crossroads and it was chosen to lead a radiobiological survey of Bikini detonations after the tests. With funding from the AEC and cooperation from the United States Navy, the AFL worked on monitoring throughout the bomb testing at the Pacific Proving Grounds. Donaldson (1950) described the objectives of the monitoring program as it evolved early in the work as follows:

> The emphasis of the studies has changed from year to year as various problems and points of interest developed, but the major effort has been directed toward gathering additional information on the following basic problems:
>
> 1 The measurement of the presence or absence of radiation in the area studied.
> 2 The determination of the presence or absence of radiation in the various native faunal and floral systems.
> 3 The selection by the various organisms of radioactive materials.
> 4 Tissue selection of radioactive materials.
> 5 The effect upon biotic forms from exposure to or absorption of radioactive materials.
> 6 The role of the native fauna and flora in the translocation and concentration of radioactive materials.

These objectives would expand even further after the Bravo test in 1954 when the concern about the contamination of Rongelap Atoll changed priorities of the monitoring program.

Initially the main focus of the AFL's monitoring was on the marine biota of the coral reef at Bikini and Eniwetok. Very little was known about the biology and ecology of coral reefs in the late 1940s, so the fieldwork was largely descriptive. Eventually, as the AFL personnel participated in more research expeditions to study the effects of bomb testing, they developed knowledge about these complex ecosystems. Donaldson's (1960) description of the reefs at the proving grounds provides a vivid image of the environment they were challenged to study:

> On the reef and in the lagoon there is an abundance of colorful plant and animal life in which the keen competition between different species for space and food is very evident. On every hand there is evidence of rapid growth and simultaneous destruction. Masses of reef-building coral are competing with each other and with the coralline marine algae for space, one often overgrowing the other. Schools of green parrotfish gnaw wide scratches on the coral. Fleshy patches of algae are pressed tightly against the surface of the coral to hold against the surges of the water pushed across the reef by the crashing breakers. Sea urchins and clams grind niches into the hard coral, some of them constantly

> feeding on the cover of bacterial and algal film which is as constantly being replaced. The clams, the corals, some small fish and other forms are ceaselessly removing from suspension in the water the small, often microscopic, plants, animals and bits of debris which make up the plankton. In regions of quieter water, where sand has been deposited, sea cucumbers and spider snails, among the larger forms, turn the sand again and again in their gleaning for food.
>
> Large schools of goatfish, mullet, surgeonfish, and other plant and plankton feeders are a common sight. Preying on unwary or disabled members of these schools are the carnivorous fish—the groupers, tuna, jacks and sharks. Ultimately the waste products and carcasses of these and other carnivores are returned to the lagoon and reef to complete the cycle.
>
> Little or no time is lost between steps in the biological cycling of materials for there is not only an abundance of organisms but also a wide variety of species . . . so that whatever is not utilized by one is quickly taken by another. There is here a perfect economy of use of substances essential to life.

Their overall approach during a monitoring expedition was to sample as many species as possible, preserve them for later study or process them on site for later radioanalysis that was carried out at the University of Washington campus after the expedition. Terrestrial sampling was also done with measurements of radioactivity levels using hand-held Geiger counters. One general strategy was to try to examine how radioactivity was distributed across food chains of organisms, but the pathways of energy flow were poorly known during the early years of the monitoring program. Eventually, comparative studies were done before and after bomb tests at single sites, and later studies were done at a distant atoll that had not been disturbed by direct blasts or fallout.

Results of the monitoring expeditions comprised large inventories of activity levels of different radionuclides from the bomb tests in different organisms at different locations. For example, Welander (1969) lists activities of 10 radionuclides from 13 compartments (including algae, fish, seabirds, soils, land plants, rats, shorebirds and others) from 15 islands at Bikini and Eniwetok Atolls, just from one expedition in August of 1964! Useful results were derived from these inventories: for example, Welander (1969) noted that Co-60 was detected most often in marine organisms while Cs-137 and Sr-90 were detected most often in plants and animals on land; but, as might be expected, there was much variability in the data and little synthesis.

Other, more focused studies on individual species or groups of species showed, for example, thyroid injury in fishes from internal doses of Iodine-131 (Gorbman and James 1963), uptake and dissipation of radiation by land crabs (Held 1960, Chakravarti and Eisler 1961) and high concentration of Co-60 by killer clams (Weiss and Shipman 1957). The wide range of these focused studies reflect the diversity of the coral reef/atoll ecosystem.

One example of synthesis from the AFL was an early version of the rating of radiosensitivity of different groups of aquatic organisms (Table 3.1) given by

TABLE 3.1 Radiosensitivity of different groups of aquatic organisms, in terms of the dose to cause 50% mortality of the test population (Donaldson and Foster 1957).

Taxonomic Group	*Dose, Roentgens*
Algae	8,000–100,000
Protozoa	10,000–300,000
Mollusca	5,000–20,000
Crustacea	500–90,000
Fish	600–3,000

Donaldson and Foster (1957) based on work from the Pacific Proving Grounds and other research. Radiosensitivity was an important topic throughout the history of Radioecology and the data given in the table was one of the first of this kind of synthesis. A conclusion of the monitoring work done by the AFL and others was the general lack of long-term effect from the testing on aquatic and terrestrial ecosystems. Of course, short-term effects of testing were dramatic, with complete devastation of sections of the reef by the explosion of a bomb, but long-term effects, as might be revealed by resurveys done a year or more after a detonation, were often not apparent. For example, Blinks (1952) surveyed marine algae one year after Operation Crossroads at Bikini and found a few examples of algae with elevated radioactivity levels (such as in *Halimeda* and *Valonia*) but complete recolonization of previously devastated substrates with representative species.

Long-term monitoring of land plants after the detonation code-named Nectar in 1954 on Eniwetok was described by Palumbo (1962). The study site was located 4.3 kilometers (2.7 miles) from the location of the explosion on one of the small islands of the atoll. Recovery of tagged plants was monitored over a three-year period after stress from radiation, shock and heat blast from the detonation. Most species regrew rapidly after the explosion, as illustrated in a time series of photographs given the in the publication. Essentially no abnormalities were observed in this study. Earlier studies of other detonations in other locations had reported some abnormalities in plants. For example, Fosberg (1959) reported "chlorosis, dead terminal twigs and mistletoe-like abnormal growths" in plants from Rongelap Atoll. Palumbo (1962) discussed these observations and suggested that they may have been caused by effects other than radiation, such as nutrient deficiency, drought or other causes, based on his fieldwork.

One of the most remarkable stories of the monitoring program in the Pacific Proving Grounds involved the rats of Engebi, one of the islands in the Eniwetok Atoll. The population of roof rats (*Rattus rattus*) on this island survived and repopulated terrestrial habitats after multiple nuclear detonations on or near the island. These detonations included the hydrogen bomb code-named Mike, which was rated at 10.4 megatons, the largest test done at Eniwetok. Roof rats were introduced to the atoll probably by the Japanese occupation during World War II, and they coexisted with the Polynesian rat (*Rattus exulans*), which had been introduced earlier by the native Micronesian people. These rats are primarily herbivorous, though the roof rat utilized a wider range of foods, including insects (Fall et al. 1971). The Polynesian rats seem to have been exterminated by detonation impacts, but the roof rats live underground in burrows and this feature probably helped them survive the blast, radiation and tidal waves caused by the bomb detonations. W. B. Jackson (1969) described the survival of rats and corrected earlier taxonomic confusion between the species. Jackson had extensive experience with Pacific rat biology, having been a member of the "PIRATE" (Pacific Island Rat Ecology) project (Storer 1962). As an aside, he also later participated in the El Verde project in Puerto Rico (see Chapter 7). Jackson and other workers (Jackson and Carpenter 1969, Bastian and Jackson 1976, Welander 1969) found that rats concentrated Cs-137 from the short food chains on the test islands but no abnormalities were identified during the test period. Later studies suggested some longer-term genetic changes may have occurred, but these did not seem to affect the biology of the rats (Jackson et al. 1987). The survival of roof rats after nuclear detonations was facilitated by their well-known adaptability and their fast reproductive rates, along with their burrowing behavior. These qualities pre-adapt them as urban pests, but in the case of the Pacific Proving Grounds, they also allowed the rats to contribute to the self-organization of the terrestrial ecosystems on the atolls following bomb testing.

After the disaster of the Bravo test at Bikini in 1954, the priorities of the monitoring program expanded to cover an evaluation of the residual fallout radiation at Rongelap. Specifically, the AFL was tasked with evaluating whether and when the people who had been evacuated from the island could return, based on radioactivity levels in their food sources, both terrestrial and marine (see, for example, Held 1961). More work was undertaken on the terrestrial ecosystem, which stretched the expertise of the AFL researchers, and scientists with expertise in soil science were recruited from the University of Washington for additional studies (Walker et al. 1961, Cole et al. 1961). One interesting issue involved the high concentration of Cs-137 in the plants at Rongelap. The potassium level in the atoll soils was found to be low and perhaps a limiting factor, so it was suspected that Cs-137 was being taken up by the plants in greater quantities than would otherwise be the case. The suggestion was made to amend the soils with potassium to attempt to reduce Cs-137 taken up by food plants of the local people (Walker et al. 1961). Cohn et al. (1960) reviewed radioactivity levels two years after Bravo in water, soil, plants, fishes and land animals and included assessments of exposures to people who would utilize the food sources on the island. Some important foods, including coconut palms, crabs and clams, were found to have high concentrations of multiple fission products. There was no precedent for predicting how humans would accumulate radiation in their foods and, interestingly, it was suggested that the surviving rat populations on Rongelap might serve as a model for the body burdens of radiation that humans feeding from the island food chains might develop! Ultimately, radiological assessments concerning resettlement of Rongelap continued for many years with little resolution (NRC 1994).

Another issue that stretched the capabilities of the AFL was the need to monitor the open ocean ecosystem that had been contaminated by Bravo and the other bomb testing. This was necessary because fisheries were involved, which might concentrate fallout elements and thus have an impact on humans who ate the harvested species. Hines (1962) describes this issue in particular regarding the concern of Japan, from which ships fished the area in and around the Pacific Proving Grounds. The AFL with the help of the U.S. Navy conducted ocean cruises sampling radioactivity levels in water, plankton and fishes in attempts to monitor this hazard. Lowman (1960, 1963) describes some data from this open ocean monitoring work. Data included radioactivity levels in organs and meat of tuna that were caught and analyzed during the ocean cruises.

In overview the environmental monitoring work done in the Pacific by the AFL and others was heroic. With little background in tropical marine systems they accomplished their objectives of evaluating the distribution of radioactivity in aquatic and terrestrial organisms. They had many challenges in conducting research in the remote location of the Pacific Proving Grounds. For example, while the U.S. Navy seems to have been quite helpful (Hines 1962), the scientists certainly had to adapt their research to the availability of ships and small boats. They were typically not given information about the details of the detonations, such as the type of device or the magnitude of the explosion, and funding from the AEC for the research was limited, especially before the post-Bravo studies done at Rongelap in regard to the relocation of the residents who had been evacuated from the atoll. Interviews conducted by Rothschild (2013) record these and other types of dissatisfaction that the researchers had about the monitoring program. Donaldson and his colleagues deserve a great deal of credit for designing a research program to describe the impacts of bomb testing on the complex, shallow-water coral reefs on one hand and the vast, deep-water open ocean that surrounded the atolls on the other hand. They were largely unprepared for the work but they did the best they could under challenging circumstances.

In retrospect, the data gathered during the bomb testing at the Pacific Proving Grounds had little impact on the science of Ecology. In part due to national security concerns, much of the data was never used or synthesized and it remains in unpublished contract reports prepared for the AEC. The monitoring work was done at the beginning of Radioecology, in the late 1940s and early 1950s, and in a sense it initiated the field of study. During this period the ecosystem concept was operationally a new model for organizing research and for synthesizing data and it was not really used by scientists involved with monitoring the bomb testing. Figure 3.1 is the only published ecosystem diagram produced from data gathered in the monitoring program. One interesting aspect of atoll ecology that was discovered during the monitoring program was the special role of terns in cycling radionuclides. These birds feed on fish in the waters surrounding the atoll but nest on the terrestrial island and, therefore, they are shown in Figure

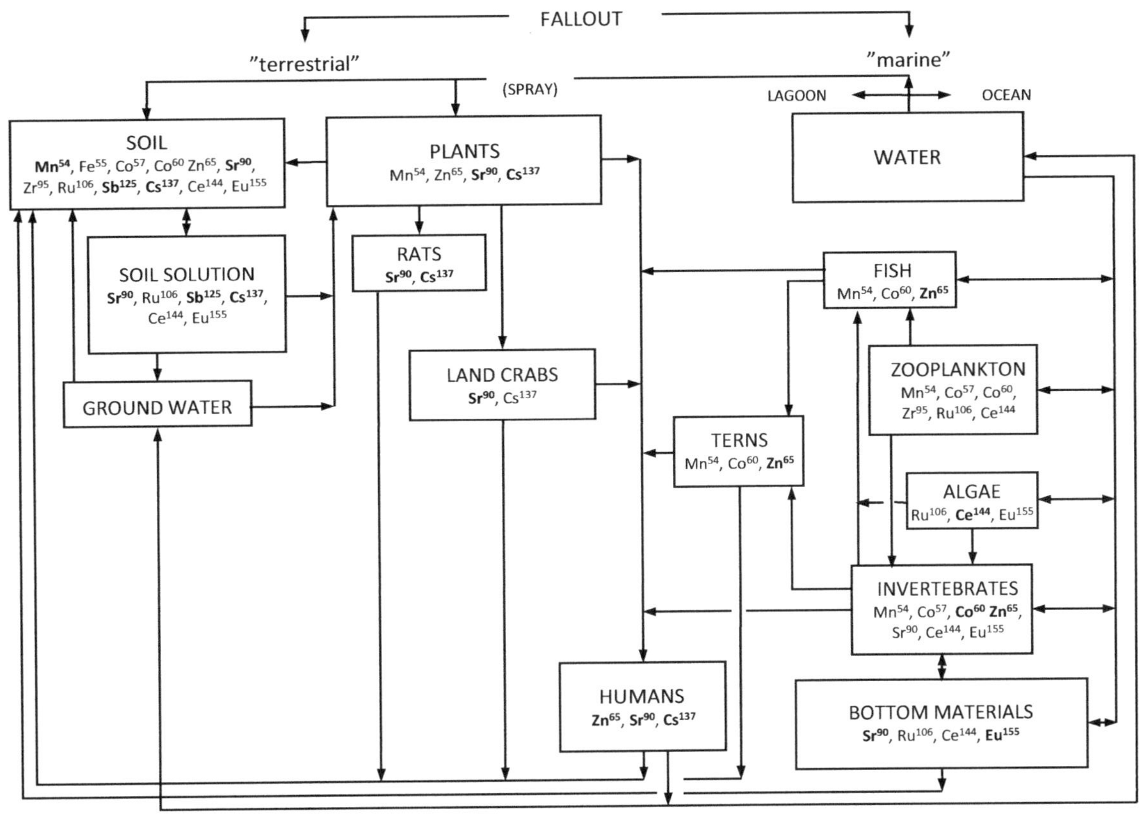

FIGURE 3.1 Ecosystem compartment model for the Pacific coral reefs studied by the University of Washington Applied Fisheries Laboratory at the Pacific Proving Grounds (from Hines 1962).

Source: Used with permission of the University of Washington Press.

3.1 as connecting the "marine" system with the "terrestrial" system. The discovery of this circuit in the mineral cycle of the atolls was an early contribution of the monitoring program to the development of a systems understanding of the atoll environment that would grow out of future research at the Pacific Proving Grounds. Later studies have shown this particular pathway to be significant in nutrient cycling on coral islands (Allaway and Ashford 1984).

Ecosystem Analysis at Eniwetok

An important study conducted in the Pacific Proving Grounds had very little to do with Radioecology directly, but the work left a lasting imprint in the field of Ecology as a whole. H. T. Odum and E. P. Odum were brothers who had already established strong academic careers before their classic study of coral reef ecology at Eniwetok in 1954 (Odum and Odum 1955). They came to work at the newly opened Eniwetok Marine Field Station for a six-week period during July and August 1954 with support from the AEC. This period was immediately after the detonation of the first hydrogen bomb (code name Mike) at Eniwetok at the small island of Elugelap, which was completely obliterated by the explosion. The Odum brothers were relatively young and very energetic: H. T. Odum was 30 years old, and E. P. Odum was 41 years old, and they carried out extensive research on the coral reef during their brief stay on the atoll. The first edition of E. P. Odum's important textbook *Fundamentals of Ecology* had just been published in 1953. (H. T. Odum had made a special contribution to the chapter on energy flow.) Their AEC-supported work at Eniwetok was to field-test some of the concepts on the coral reef where there were open questions about ecological energetics. After their research at Eniwetok, neither brother would ever do significant fieldwork on coral reefs again during their long careers, but their time spent on the atoll built a foundation for future work in other ecosystem types.

They chose to study a part of the atoll that had not been directly affected by testing: the reef adjacent to Japtan Island. Overall their research involved assessing biomass and metabolism of the coral reef, in other words structure and function of the reef ecosystem. Metabolism, in terms of net daytime photosynthesis (P) and nighttime respiration (R) was measured with the upstream-downstream methods of change in dissolved oxygen concentration in the water as it passed from the ocean over the reef and into the lagoon. They chose a location with unidirectional water flow across the reef so this method could be applied. This is a convenient method for measuring metabolism in aquatic ecosystems and H. T. Odum had experience with it from earlier work in freshwater spring runs in Florida.

In their studies of diurnal oxygen changes at Eniwetok, the Odum brothers closely compared their data with measurements made nearly a decade earlier at Rongelap Atoll by Sargent and Austin (1949). This earlier study was important to the Odums who cited it 13 times in their paper (Odum and Odum 1955). Both the measurements at Rongelap in 1946 by Sargent and Austin and those done by the Odums at Eniwetok demonstrated that the coral reefs had much higher metabolism than the surrounding ocean waters and the rates of biological production (P) and consumption (R) were nearly balanced for the reef ecosystems, suggesting "a true ecological climax or open steady-state system" (Odum and Odum 1955).

Sampling and estimating biomass of the living populations took more time and effort than the metabolism measurements due to the diversity of species and habitats of the reef. The reef has a zonation pattern of habitats along the unidirectional water flow pathway from the ocean to the lagoon, primarily determined by the physical energies of waves, currents and tides. Thus, the Odums established a transect of quadrats, each 6 meters x 6 meters (20 feet x 20 feet), along the water flow pathway with one quadrat per zone. They mapped the cover types (Figure 3.5 in their paper) and estimated biomass of the living populations (see their Tables 8–12) in the quadrats. Organisms were categorized by trophic level (e.g., primary producers, herbivores and carnivores) as part of the study. A variety of methods was required to sample the organisms since the diversity spanned the tree of life, from microbes to invertebrates and macroalgae to free-ranging large fishes. Example of the sampling challenges the Odums faced are a method they developed using melted Vaseline to estimate coral polyp biomass (p. 305) and a visual survey method to estimate shark biomass (p. 309)! The pyramids of biomass that resulted from their studies (Figure 3.2) illustrate the classic pattern of declining biomass across the trophic levels, from producers to herbivores to carnivores, caused by energy losses at each step due to the Second Law of Thermodynamics. The average biomass across all the quadrats in dry weight was 703 grams/m^2 for producers, 132 grams/m^2 for herbivores and 11 grams/m^2 for carnivores (shark biomass was 1.2 grams/m^2).

One of the details of the biomass inventory was the dominance of algae on the reef. As a consequence, as has been noted, referring to the ecosystem as a coral reef is a bit of misnomer—it might better be described as an algal reef. A variety of algal groups were inventoried by the Odums including zooxanthellae, or the single-celled dinoflagellates found in the coral polyp tissue. In the early 1950s these algae were thought by most scientists to be symbiotic with the corals: the algae utilizing nutrients released from respiration by the coral animal and carrying out photosynthesis with carbon molecules that are present, providing energy for the coral. This symbiotic pairing allows for an effective cycling of nutrients on the reef, which is important since nutrient concentrations in the oceanic water surrounding the reef are low. The Odums estimated zooxanthellae biomass and found it to be only 6% of the total algal biomass on the reef. A larger mass occurred in the filamentous algae found throughout the calcium carbonate coral skeleton. The Odums portrayed the relative distributions of algae in a cross-section diagram which matches the biomass values of the different components with their locations in the coral-algal system. The form of the diagram has become a standard way of illustrating the coral polyp-algal symbiosis (see, for example, Hatcher 1988) and E. P. Odum replicated it in subsequent editions of his text (Odum 1959, 1971). The Odums speculated that these free-living filamentous algae were also symbiotic with the coral. They made no direct measurement of exchanges with the symbiotic pairing but they based the inference on radioaudiographs of cross-sections through living

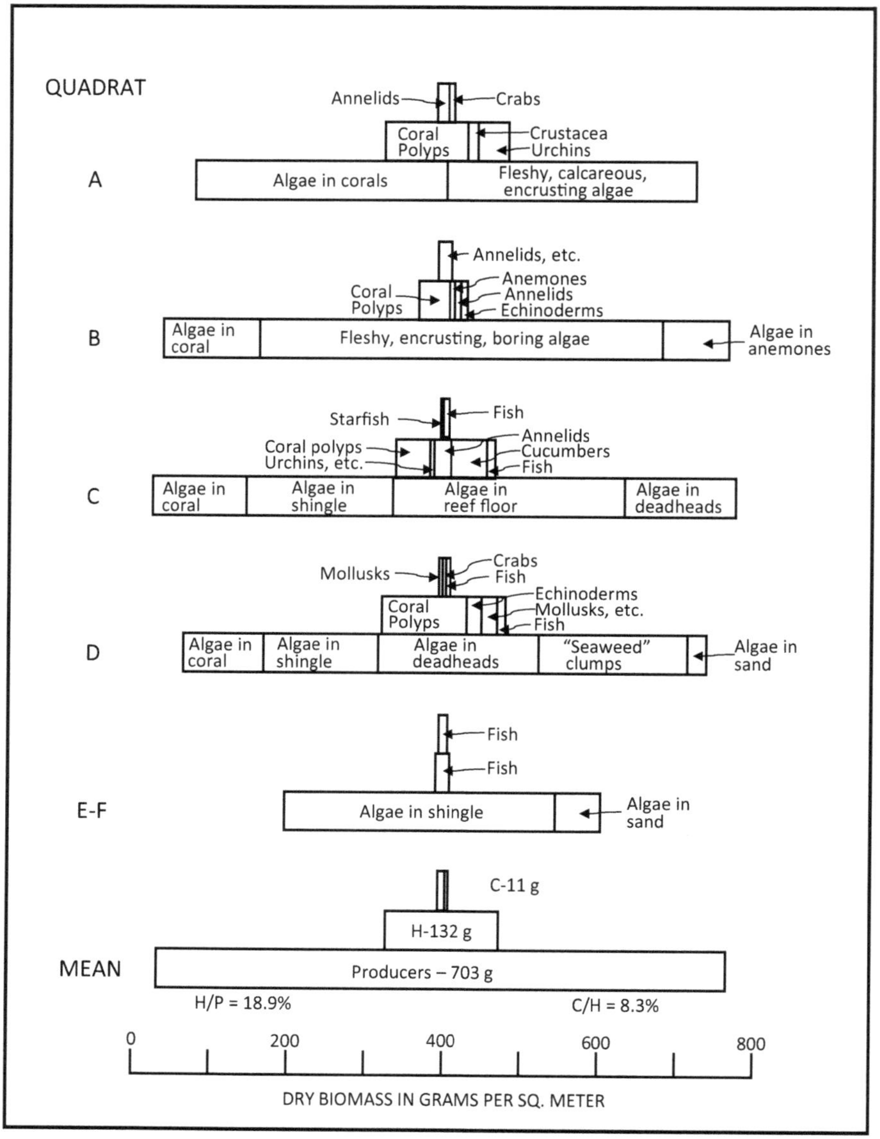

FIGURE 3.2 Pyramids of biomass found along the transects studied by Odum and Odum (1955). For each pyramid, producers are shown on the bottom, herbivores are shown in the middle and carnivores are shown on the top.

Source: Used with permission of John Wiley & Sons.

and dead coral skeletons. The algae filaments in dead coral skeletons produced bright radioaudiographs indicating high levels of radioactivity. This suggests that these algae had taken up fallout radionuclides from the surrounding water. However, the algae filaments associated with living coral colonies showed no evidence of radioactivity in the radioaudiographs. This suggested to the Odums that these algae were not taking up nutrients from the surrounding waters but rather were receiving them from the coral polyp in a symbiotic exchange.

E. P. Odum was especially impressed by their studies of the role of coral-algal symbiosis in the functioning of the reef, and 20 years later he referred to it as an emergent property of the ecosystem (Odum 1977). He went further to wonder about a possible lesson for human systems, as can be seen in the following quote:

> Do these coral reef discoveries have any significance for urban industrial man? Perhaps they do. The Pacific coral reef, as a kind of oasis in the desert, can stand as an object lesson for man who must now learn that mutualism between autotrophic and heterotrophic components, and between producers and consumers in the societal realm, coupled with efficient recycling of materials and use of energy, are the keys to maintaining prosperity in a world of limited resources.
>
> *(Odum 1977)*

Although some subsequent studies of coral reef algae have suggested a smaller role for the free-living filaments in the coral skeletons (Schlichter et al. 1997, Jacques and Pilson 1980), the importance of symbiosis between corals and algae is an accepted fact, which as an aside was later verified empirically with radioisotope experiments, as reviewed by Bowen (2015).

The Odums didn't have enough time to cover all of the important aspects of reef ecology during their six-week stay. For example, decomposers were not studied, and they stated, "An understanding of the trophic relationships of the bacteria and foraminifera on reefs is urgently needed" (Odum and Odum 1955). They also later wondered about the role of bacteria in the coral heads or skeletons (Odum and Odum 1956). H. T. Odum supervised a doctoral student, Louis DiSalvo, 15 years later who studied this aspect of reef ecology (DiSalvo 1969) with fieldwork in Hawaii and at Eniwetok. DiSalvo (1969, 1971) found bacterial counts of 10^4 to 10^5 bacteria per gram dry weight of coral skeletons with high oxygen consumption rates suggesting significant nutrient regeneration activity. This kind of research has continued to identify nutrient cycling pathways in the complexity of the coral reef (Mumby and Steneck 2018, Knowlton and Rohwer 2003).

In summary, the research conducted by the Odums at Eniwetok was highly regarded in the field of Ecology. They won the George Mercer Award from the Ecological Society of America in 1956 (Anonymous 1957). Their work has been variously termed "a significant contribution" (Hand 1956), a "pioneer work on the metabolism of communities" (Sorokin 1995), "a landmark in ecological research" (Hagan 1992), a "foundation" study (Smith 1985), a "classic study" (Karlson 1999) and "an astonishing accomplishment" (Barile 2004). Kaplan (1982) went ever further when he claimed the Odums' research "produced some of the most challenging and creative hypotheses concerning the coral reef since Darwin".

Heritage

Bomb testing at the Pacific Proving Grounds ended in 1958, but studies of the effects of irradiating the atolls continued for decades. A special issue of the journal *Health Physics* (Volume 73, Number 1) provides an overview of the research done during the testing years and afterwards (Simon and Vetter 1997). Many studies were needed after tests had ended to determine if and when the original inhabitants could safely return to their homeland and what remediation of the radioactive contamination was needed. After the detonations at Bikini and Eniwetok there was much residual radioactivity on the atolls. An extreme example is described for one of the islands on Eniwetok by Wargo:

> On Runit Island, which lies with the atoll, a 15-kiloton bomb exploded but did not produce a nuclear blast, spreading plutonium over much of the island site. Radiation levels are expected to preclude resettlement for at least 24,000 years. The Defense Nuclear Agency, recognizing the intensity of the problem, decided not to clean it up, and instead made the island a repository for the most highly contaminated debris removed from other nearby islands. The crater left in the lagoon and the dome constructed above the plutonium waste are easily visible in public-domain satellite images of the area.
>
> In creating the nuclear waste depository, nearly 100,000 cubic yards of radioactive soils were scraped from the islands, mixed with concrete, and placed within an atomic bomb crater on Runit Island.
>
> *(2009)*

Restoration by the AEC included plowing the soils, removing vegetation and replanting coconut trees and other crops to encourage resettlement at a cost of more than $100 million (Wargo 2009). After several false starts, some of the people have returned, but questions about safety at the atolls continue to the present day.

However, developments from work done during the testing period of 1946 to 1958 do provide heritage of this early phase in the history of Radioecology. Some of the scientists involved in research at the proving grounds went on to lead other projects in Radioecology. Of course, the Odum brothers were important at future works: E. P. Odum at the Savannah River Ecology Laboratory (see Chapter 9) and H. T. Odum at El Verde, Puerto Rico (see Chapter 7). Also, Lauren Donaldson, who had led the monitoring research throughout the bomb testing period, went on to attempt to use radioactive tracers as part of a research program for improving salmon fisheries at his home institution, the University of Washington. The AEC provided support for his studies in a coastal freshwater lake but, due to several unanticipated limitations, the research did not lead to technology for increasing salmon production (Klingle 1998).

Because of the circumstances of monitoring research, especially the remote location and security issues, there were little or no opportunities for graduate student training at the Pacific Proving Grounds. One notable exception was Thomas Goreau, who participated as a young chemist during the radiological monitoring for Operation Crossroads at Bikini (Bowen 2015). Goreau went on to become a leading coral reef ecologist who, among other accomplishments, was involved with using radioactive tracers to study the role of zooxanthellae in the coral-algal symbiosis (Goreau and Bowen 1955, Goreau and Goreau 1960).

Perhaps the most significant academic heritage of Radioecology at the Pacific Proving Grounds had more to do with geology than Ecology. In preparation for bomb testing, the AEC commissioned deep drilling of the limestone of Bikini and Eniwetok Atolls to help determine whether they could withstand the explosive power of the detonations. Drilling was undertaken by a team of geologists and work extended over a number of years (Ladd and Schlanger 1960). The drilling efforts eventually reached depths of more than 1200 meters (4000 feet). Most of the depth of the bores was composed of limestone deposited by the calcareous algae and corals over hundreds of thousands of years. However, one of the deepest bores from the island of Elugelab on the Eniwetok Atoll yielded basalt, representing a volcanic foundation for the coral reef. This finding gave confidence that the island could withstand the blast from the detonations and it also was relevant to the long-standing controversy about the origins of the Pacific Ocean atolls (Dobbs 2005, Davis 1928). Several rival theories had been proposed by geologist-explorers through the 1800s and 1900s. One was Charles Darwin's subsidence theory, based on observations he made during the voyage of the H.M.S. *Beagle* (Rosen 1982). Darwin proposed that the main kinds of coral reefs—fringing, barrier and atolls (like Bikini and Eniwetok)—formed a developmental sequence in relation to eroding volcanic islands. The sequence begins with fringing reefs, which are found in shallow water attached to the newly formed volcanic islands. As the island erodes over geologic time, the reef continues to grow upward, passing through different phases: as a barrier reef separated from the island by a channel and finally as a free-standing, circular atoll after the original volcano has eroded below the surface of the ocean. Thus, the finding of basalt from a volcano at depth in the Eniwetok bore holes provided empirical support for Darwin's theory. There was irony in this finding, however, in that the island, where the drilling had yielded the basaltic foundation and where Darwin's theory was confirmed, was obliterated by the detonation of the first hydrogen bomb (code name Mike), leaving a massive hole in the circular atoll! Davidson (1998) provides a dramatic description of what happened to the island that had proven Darwin's theory:

> At a fraction of a second before 7:15 A.M. on the first of November, 1952, Mike was detonated. It exploded with a force nearly one thousand times greater than that of the atom bomb dropped on Hiroshima. The complex chain reaction begun inside of Mike resulted in a fireball that burned hotter than the sun.
>
> It was dawn. Coral polyps, fat with zoox, would have been slowly withdrawing into their calcium dens after a nighttime spent capturing

> zooplankton. Parrotfish would have been emerging from their mucus cocoons to begin their daily hunt for algae. Crabs, shrimp, and other crustaceans would have been crawling from their lairs. It was that magical moment of transition, with creatures of the sunlight replacing the many nocturnal feeders.
>
> In nanoseconds they were all vaporized, transformed into a swirling, purplish cloud of unimaginably hot gases. This fiery mist shot twenty-seven miles up into the pale blue Pacific dawn, where it drifted with the cold winds blowing along the edge of the Earth's atmosphere.

A number of general studies of flora and fauna were completed as part of the AEC-funded monitoring program that provide ongoing value as heritage of Radioecology on the Pacific Proving Grounds. These included surveys of algae (Gilmartin 1960), terrestrial plants (Taylor 1950, St. John 1960) and many other groups. The biota of Pacific Ocean atolls was little known before the atomic bomb testing period, so these biological surveys provide a foundation for future research. Perhaps the most significant example of this kind of work was the review of the fish fauna by Hiatt and Strasburg (1960) that extended from after Operation Crossroads in 1947 to 1955. They studied the food habits of 233 species of coral reef fishes and provided trophic classifications of all the species based on stomach content analysis. This work was an asset to the Odum brothers during their ecosystem analysis at Eniwetok and it continues to be an important resource for Pacific coral reef biologists.

An interesting set of studies of Pacific atolls was undertaken in the years 1950–1954 by the Pacific Science Board, which is a branch of the National Research Council of the National Academy of Sciences. This was a broad survey of the geography and natural history and human ecology of four atolls. It was funded by the U.S. Office of Naval Research "to make an effort to fill the gap in scientific knowledge about Pacific coral reefs and atolls" (Wiens 1962). The project resulted in a number of publications, including syntheses by Tracey et al. (1961), Bates and Abbott (1958) and Wiens (1962). The intensive work by Niering (1963) is an excellent description of the terrestrial ecology of Kapingamarangi Atoll, which includes an energy flow diagram that was probably representative of the ecosystems from the atolls of the Pacific Proving Grounds. This interesting scientific program does not seem to have been described from a historical perspective and it would be useful to know if it had a direct connection to the AEC's work with the Pacific atolls.

An important element of heritage of the Radioecology research of the Pacific Proving Grounds was the application of the ecosystem concept to coral reefs. The Odum brothers' contribution in 1954 was the foundation for this way of thinking. Their work was built upon 17 years later by an interdisciplinary group of scientists who visited Eniwetok to reexamine the earlier research with more sophisticated methods. This reexamination occurred in Spring 1971 when 25 ecologists and oceanographers carried out the two-month-long Symbios expedition to Eniwetok with the research vessel *Alpha Helix* from the Scripps Institution of Oceanography

(D'Elia and Harris 2008). The expedition was supported by the AEC and the NSF and it produced detailed estimates of metabolism and nutrient cycling for the coral reef in several locations, which included the Odums' original research transect (Johannes et al. 1972). The work of the Symbios expedition largely confirmed the findings of the Odum brothers with a near balance of production and respiration and efficient cycling of phosphorus. New measurements demonstrated the importance of nitrogen fixation, which was suggested to be occurring "at rates of the same high order as those reported for alfalfa fields" (Johannes et al. 1972). The work included the first detailed carbonate budget for a coral reef (Smith 1973), along with many other studies. Sapp (1999) called the expedition "a major turning point in coral-reef research. It clearly demonstrated how one could study a reef as a system, not just observing species, but examining energy flows."

As an aside, the principal leader of the Symbios expedition was R. E. Johannes, whose doctoral work (Johannes 1964) had been supported by the AEC. He would later go on to describe the sustainable balance of an indigenous fishing culture of another Pacific atoll (Johannes 1981), perhaps illustrating some of the lessons of the climax coral reef ecosystem for human societies that the Odum brothers had foretold in their research during the period of bomb testing at Eniwetok.

Literature Cited

Allaway, W. G. and A. E. Ashford. 1984. Nutrient input by seabirds to the forest on a coral island of the Great Barrier Reef. Marine Ecology—Progress Series 19:297–298.

Anonymous. 1957. The George Mercer Award for 1956. Bulletin of the Ecological Society of America 38:2–3.

Barile, P. J. 2004. Comments on "Trophic structure and productivity of a windward coral reef community on Eniwetok Atoll" [Ecological Monographs 25(3)(1955)291–320]. Ecological Modelling 178:51–57.

Bastian, R. K. and W. B. Jackson. 1976. ^{137}Cs and ^{60}Co in a terrestrial community at Enewetak Atoll. Pp. 314–320. In: C. E. Cushing, Jr. (ed.). Radioecology and Energy Resources. Dowden, Hutchinson & Ross, Inc., Stroudsburg, PA.

Bates, M. and D. P. Abbott. 1958. Coral Island, Portrait of an Atoll. Charles Scribner's Sons, New York, NY.

Blinks, L. R. 1952. Effects of radiation on marine algae. Journal of Cellular and Comparative Physiology 39(supplement 2):11–18.

Bowen, J. 2015. The Coral Reef Era: From Discovery to Decline, A History of Scientific Investigation from 1600 to the Anthropocene Epoch. Springer, New York, NY.

Chakravarti, D. and R. Eisler. 1961. Strontium-90 and gross beta activity in the fat and nonfat fractions of the liver of the coconut crab (*Birgus latro*) Collected at Rongelap Atoll during March 1958. Pacific Science 15:155–159.

Cohn, S. H., J. S. Robertson and R. A. Conard. 1960. Radioisotopes and environmental circumstances: The internal radioactive contamination of a Pacific Island community exposed to local fallout. Pp. 306–330. In: R. S. Caldecott and L. A. Snyder (eds.). Radioisotopes in the Biosphere. University of Minnesota, Minneapolis, MN.

Cole, D. W., S. P. Gessel and E. E. Held. 1961. Tension lysimeter studies of ion and moisture movement in glacial till and coral atoll soils. Soil Science Society of America Proceedings 25:321–325.

Davidson, O. G. 1998. The Enchanted Braid: Coming to Terms with Nature on the Coral Reef. John Wiley & Sons, New York, NY.
Davis, W. M. 1928. The Coral Reef Problem. Special Publication No. 9. American Geographical Society, New York, NY.
D'Elia, C. R. and A. R. Harris. 2008. The R/V Alpha Helix expedition: A retrospective analysis of a milestone in coral reef research. pp. 38–42. In: Proceedings of the 11th International Coral Reef Symposium. Ft. Lauderdale, FL.
Dibblin, J. 1988. Day of Two Suns, U.S. Nuclear Testing and the Pacific Islanders. New Amsterdam Books, New York, NY.
DiSalvo, L. H. 1969. Isolation of bacteria from the corralum of *Porites lobata* (Vaughan) and its possible significance. American Zoologist 9:735–740.
DiSalvo, L. H. 1971. Regenerative functions and microbial ecology of coral reefs: labeled bacteria in a coral reef microcosm. Journal of Experimental Marine Biology and Ecology 7:123–136.
Dobbs, D. 2005. Reef Madness, Charles Darwin, Alexander Agassiz, and the Meaning of Coral. Pantheon Books, New York, NY.
Donaldson, L. R. 1950. Radiobiological Survey of Bikini, Eniwetok, and Likiep Atolls July-August 1949. AECD-3446, U.S. Atomic Energy Commission, Washington, DC.
Donaldson, L. R. 1960. Radiobiological studies at the Eniwetok test site and adjacent areas of the western Pacific. Pp. 1–7. In: C. M. Tarzwell (ed.). Biological Problems in Water Pollution. U.S. Public Health Service, Robert A. Taft Sanitary Engineering Center, Cincinnati, OH.
Donaldson, L. R. and R. F. Foster. 1957. Effects of radiation on aquatic organisms. Pp. 96–102. In: The Effects of Atomic Radiation on Oceanography and Fisheries. Publication No. 551. National Academy of Sciences, Washington, DC.
Fall, M. W., A. B. Medina and W. B. Jackson. 1971. Feeding patterns of *Rattus rattus* and *Rattus exulans* on Eniwetok Atoll, Marshall Islands. Journal of Mammalogy 52:69–76.
Fosberg, F. R. 1959. Plants and fall-out. Nature 183:1448.
Gilmartin, M. 1960. The ecological distribution of the deep water algae of Eniwetok Atoll. Ecology 41:210–221.
Gorbman, A. and M. S. James. 1963. An exploratory study of radiation damage in the thyroids of coral reef fishes from the Eniwetok Atoll. Pp. 385–399. In: V. Schultz and A. W. Klement, Jr. (eds.). Radioecology, Proceedings of the First National Symposium on Radioecology. Reinhold Publ. Corp., New York, NY.
Goreau, T. F. and V. T. Bowen. 1955. Calcium uptake by a coral. Science 122:1188–1189.
Goreau, T. F. and N. I. Goreau. 1960. Distribution of labeled carbon in reef-building corals with and without zooxanthellae. Science 131:668–669.
Hagan, J. B. 1992. An Entangled Bank: The Origins of Ecosystem Ecology. Rutgers University Press, New Brunswick, NJ.
Hand, C. 1956. Are corals really herbivores. Ecology 37:384–385.
Hatcher, B. G. 1988. Coral reef primary productivity: A beggar's banquet. Trends in Ecology and Evolution 3:106–111.
Held, E. E. 1960. Land crabs and fission products at Eniwetok Atoll. Pacific Science 14:18–27.
Held, E. E. 1961. Qualitative distribution of radionuclides at Rongelap Atoll. Pp. 167–169. In: V. Schultz and A. W. Klement, Jr. (eds.). Radioecology, Proceedings of the First National Symposium on Radioecology. Reinhold Publ. Corp., New York, NY.
Hezel, F. X. 1995. Strangers in Their Own Land: A Century of Colonial Rule in the Caroline and Marshall Islands. University of Hawai'i Press, Honolulu, HI.

Hiatt, R. W. and D. W. Strasburg. 1960. Ecological relationships of the fish fauna on coral reefs of the Marshall Islands. Ecological Monographs 30:65–127.

Hines, N. O. 1962. Proving Ground: An Account of the Radiobiological Studies in the Pacific, 1946–1961. University of Washington Press, Seattle, WA.

Hines, N. O. 1966. Atoms, Nature, and Man. Understanding the Atom Series, U.S. Atomic Energy Commission, Division of Technical Information, Washington, DC.

Jackson, W. B. 1969. Survival of rats at Eniwetok Atoll. Pacific Science 23:265–275.

Jackson, W. B. and M. L. Carpenter. 1969. Radioisotope cycling in terrestrial communities at Eniwetok Atoll. Pp. 644–651. In: D. J. Nelson and F. C. Evans (eds.). Symposium on Radioecology, Proceedings of the Second National Symposium. CONF-670503 U.S. Atomic Energy Commission, Washington, DC.

Jackson, W. B., S. H. Vessey and R. K. Bastian. 1987. Biology of the rodents of Enewetak Atoll. Pp. 203–214. In: The Natural History of Enewetak Atoll, Volume 1. DOE/EV/00703-T1-Volume 1. U.S. Department of Energy, Washington, DC.

Jacques, T. G. and M. E. Q. Pilson. 1980. Experimental ecology of the temperate scleractinian coral *Astrangia danae* I: Partition of respiration photosynthesis and calcification between host and symbionts. Marine Biology 60:167–178.

Johannes, R. E. 1964. Uptake and release of phosphorus by a benthic marine amphipod. Limnology and Oceanography 9:235–242.

Johannes, R. E. 1981. Words of the Lagoon, Fishing and Marine Lore in the Palau District of Micronesia. University of California Press, Berkeley, CA.

Johannes, R. E. and 25 co-authors. 1972. The metabolism of some coral reef communities: A team study of nutrient and energy flux at Eniwetok. Bioscience 22:541–543.

Kaplan, E. 1982. Coral Reefs: The Peterson Field Guide Series. Houghton Mifflin, Boston, MA.

Karlson, R. H. 1999. Dynamics of Coral Communities. Kluwer Academic Publ., Dordrecht.

Kiste, R. C. 1974. The Bikinians: A Study in Forced Migration. Cummings Publ. Co., Menlo Park, CA.

Klingle, M. W. 1998. Plying atomic waters: Lauren Donaldson and the "Fern Lake Concept" of fisheries Management. Journal of the History of Biology 31:1–32.

Knowlton, N. and F. Rohwer. 2003. Multispecies microbial mutualisms on coral reefs: The host and a habitat. American Naturalist, Supplement 162:S51–S62.

Ladd, H. and S. Schlanger. 1960. Drilling Operations on Eniwetok Atoll. U.S. Geological Survey, Professional Papers 260-Y.

Lowman, F. G. 1960. Marine biological investigations at the Eniwetok test site. Pp. 106–138. In: Disposal of Radioactive Wastes II. International Atomic Energy Agency, Vienna, Austria.

Lowman, F. G. 1963. Radionuclides in plankton and tuna from the Central Pacific. Pp. 145–149. In: V. Schultz and A. W. Klement, Jr. (eds.). Radioecology, Proceedings of the First National Symposium. Reinhold Publishing Corp., New York, NY.

Martin, L. J. 2018. Proving grounds: Ecological fieldwork in the Pacific and the materialization of ecosystems. Environmental History 23:567–592.

Matashichi, O. 2011. The Day the Sun Rose in the West, Bikini, The Lucky Dragon, and I (translated by R. H. Minear). University of Hawai'i Press, Honolulu, HI.

McMahon, R. J. 2003. The Cold War, A Very Short Introduction. Oxford University Press, Oxford.

Mumby, P. J. and R. S. Steneck. 2018. Paradigm lost: Dynamic nutrients and missing detritus on coral reefs. Bioscience 68:487–495.

National Research Council. 1994. Radiological Assessments for Resettlement of Rongelap in the Republic of the Marshall Islands. National Academy Press, Washington, DC.

Niering, W. A. 1963. Terrestrial ecology of Kapingamarangi Atoll, Caroline Islands. Ecological Monographs 33:131–160.
Odum, E. P. 1959. Fundamentals of Ecology, Second Edition. W. B. Saunders Co., Philadelphia, PA.
Odum, E. P. 1971. Fundamentals of Ecology, Third Edition. W. B. Saunders Co., Philadelphia, PA.
Odum, E. P. 1977. The emergence of Ecology as a new integrative discipline. Science 195:1289–1293.
Odum, H. T. and E. P. Odum. 1955. Trophic structure and productivity of a windward coral reef community on Eniwetok Atoll. Ecological Monographs 25:291–320.
Odum, H. T. and E. P. Odum. 1956. Corals as producers, herbivores, carnivores, and possibly decomposers. Ecology 37:385.
Palumbo, R. F. 1962. Recovery of the land plants at Eniwetok Atoll following a nuclear detonation. Radiation Botany 1:182–189.
Rosen, B. R. 1982. Darwin, coral reefs, and global geology. Bioscience 32:519–525.
Rothschild, R. 2013. Environmental awareness in the Atomic Age: Radioecologists and nuclear technology. Historical Studies in the Natural Sciences 43:492–530.
Sapp, J. 1999. What Is Natural? Coral Reef Crisis. Oxford University Press, Oxford, UK.
Sargent, M. C. and T. S. Austin. 1949. Organic productivity of an atoll. American Geophysical Union Transactions 30:245–249.
Schlichter, D., H. Kampmann and S. Conrady. 1997. Trophic potential and photoecology of endolithic algae living within coral skeletons. Marine Ecology 18:299–317.
Simon, S. L. and R. J. Vetter. 1997. Editor's remarks. Health Physics 73:3.
Smith, S. V. 1973. Carbon dioxide dynamics: A record of organic carbon production, respiration, and calcification in the Eniwetok reef flat community. Limnology and Oceanography 18:106–120.
Smith, S. V. 1985. Some thoughts on the past, present, and future of studies on coral reef community metabolism. Pp. 33–34. In: M. L. Reaka (ed.). The Ecology of Coral Reefs, Volume 3, No. 1. National Oceanic and Atmospheric Administration, Washington, DC.
Sorokin, Y. I. 1995. Coral Reef Ecology. Springer, New York, NY.
St. John, H. 1960. Flora of Eniwetok Atoll. Pacific Science 14:313–336.
Storer, T. I. 1962. Pacific Island Rat Ecology. Bulletin 225, Bernice P. Bishop Museum, Honolulu, Hawaii.
Taylor, W. R. 1950. Plants of Bikini. University of Michigan Press, Ann Arbor, MI.
Tracey, J. I., Jr., D. P. Abbott and T. Arnow. 1961. The Natural History of Ifaluk Atoll, Physical Environment. Bulletin 222, Bernice P. Bishop Museum, Honolulu, HI.
Walker, R. B., E. E. Held and S. P. Gessel. 1961. Radiocesium in plants grown on Rongelap Atoll soils. In: Recent Advances in Botany Volume II. University of Toronto Press, Toronto, Canada.
Wargo, J. 2009. Green Intelligence: Creating Environments That Protect Human Health. Yale University Press, New Haven, CT.
Weiss, H. V. and W. H. Shipman. 1957. Biological concentration by killer clams of Cobalt-60 from radioactive fallout. Science 125:695.
Welander, A. D. 1969. Distribution of radionuclides in the environment of Eniwetok and Bikini Atolls, August 1964. Pp. 346–354. In: D. J. Nelson and F. C. Evans (eds.). Symposium on Radioecology, Proceedings of the Second National Symposium. CONF-67050 U.S. Atomic Energy Commission, Washington, DC.
Wiens, H. J. 1962. Atoll Environment and Ecology. Yale University Press, New Haven, CT.

4

THE NEVADA TEST SITE

Introduction

Difficulties with bomb testing at the Pacific Proving Grounds (PPG) in the Marshall Islands were apparent from the start with Operation Crossroads in 1947 and Operation Sandstone in 1948. The tests were very expensive to undertake in terms of long-distance transport of personnel and equipment and support of operations in the field. Potential continental sites for testing were studied but there was a reluctance on the part of the AEC and the military to detonate atomic weapons with the U.S. boundaries. However, this reluctance disappeared with national emergencies, which began with the advent of the first Soviet bomb test (code named "Joe 1", after Joseph Stalin) in Fall 1949. Additionally, in rapid succession the revolution in China by Mao Tse-Tung which created a new Communist world power and the initiation of the Korean War made clear the need to increase bomb testing with use of a continental site. The detonation of Joe 1 demonstrated that the Soviets had negated the U.S. superiority in atomic weapons and the military mind-set held that more testing was needed to develop more bombs and more powerful bombs. It was clear that this could not be accomplished solely at the PPG and that a continental site was necessary for more efficient and more economical testing.

A number of possible continental sites were examined, including Alamogordo/ White Sands, New Mexico; Dugway Proving Grounds, Utah; Pamlico Sound/ Camp Lejeune, North Carolina; and a site in eastern Nevada. A site in southern Nevada was ultimately chosen because of its many advantages (Titus 1986). Nuclear tests were undertaken in Alaska, as will be discussed later in the book, and a few tests (not bombs, however) were done in New Mexico, Colorado and even in a salt dome in Mississippi (Burke 2012). But the site in southern Nevada became the focus for "low-yield weapons" and other forms of nuclear testing

DOI: 10.4324/9781003284901-6

until the global ban on all testing occurred in 1991. This site, which became known as the Nevada Test Site (NTS), was initially located at an Air Force bombing and gunnery range approximately 105 kilometers (65 miles) northwest of Las Vegas. The initial site of 900 square kilometers (350 square miles) grew through the 1950s and 1960s as testing commenced, and it expanded over time to 3500 square kilometers (1350 square miles), or roughly the size of Rhode Island. Most of this land was already under federal control, which was an important advantage of the site.

The NTS was at the boundary between the Great Basin Desert to the north and the Mojave Desert to the south. The climate is hot and dry, though it varies with elevation, which ranges from about 610 meters (2000 feet) on the dry lake beds (where most of the testing took place) to the surrounding mountains, which reach to about 2225 meters (7300 feet). The weather, with low annual rainfall and predictable winds, was thought to be an advantage of the site for bomb testing, since it was believed to minimize the risks of radioactive fallout and accidents. This assessment was made by a group of experts including Edward Teller and Enrico Fermi (Fehner and Gosling 2000). Unfortunately, the weather, in particular shifting wind patterns, turned out not to be as predictable as expected and radioactive fallout reaching beyond the boundaries of the NTS became a major problem.

Because of the desert conditions and the relative remoteness of the setting, the landscape surrounding the NTS was sparsely populated. This quality of the site facilitated security and minimized the potential damage if accidents were to occur. However, despite being remote and desolate, the NTS was relatively near the city of Las Vegas, which meant that a labor force to operate the tests could commute to the site. A small town, named Mercury after a nearby abandoned mine, was developed near the highway leading to Las Vegas to house a resident staff and to serve as a headquarters for operations.

The NTS was authorized by President Truman on December 18, 1950. The intention was to restrict testing there to low-yield weapons; direct testing of the new hydrogen (i.e., fusion) bomb would take place only at the PPG, in the Marshall Islands. This original intention was not maintained and by the 1960s hydrogen devices were tested with underground detonations. Six weeks after authorization, the first tests were made at the NTS in Operation Ranger, which included five bomb detonations between January 27 and February 6, 1951. These bombs were air-dropped over Frenchman Flat but later testing was moved to Yucca Flat, further north. This site was surrounded by mountains, which provided greater safety and security for testing.

One hundred atmospheric bomb tests were carried out at the NTS from 1951 to 1963, when a test ban stopped atmospheric testing. These atmospheric tests included both air drops with detonation above the land surface and detonations made from stationary, elevated scaffolding (e.g., tower shots), both of which had advantages and disadvantages from the perspective of bomb testing (Fehner and Gosling 2000). Of particular importance, tower shots increased radioactive fallout because the rising fireball drew in large amounts of dust and debris that became coated with fission products and that was transported by wind, often off site. Fehner and Gosling (2006) provide a detailed listing of atmospheric tests held at the NTS and other sites from 1945 to 1958 in an appendix to their book.

The "Limited Test Ban Treaty" that was signed in August 1963 eliminated atmospheric testing and led to underground testing at the NTS. This move essentially eliminated radioactive fallout, though accidental venting did occasionally release fission materials into the atmosphere, such as with the Baneberry shot in 1970 (Johns and Johns 2017). Most underground testing took place in vertical shafts that were dug from 30 meters (100 feet) to more than 300 meters (1000 feet) in depth. The nuclear device and experimental equipment were lowered into the hole, which was up to 3 meters (10 feet) in diameter. The detonation vaporized the surrounding rock, creating an underground cavity below and a saucer-shaped crater of varying diameter and depth on the surface. There was little knowledge of the underlying geology at the NTS when it was established, but the AEC and the USGS supported extensive geological research once underground testing became important. This research has suggested that the geology at the NTS is appropriate for underground testing as indicated by the following quote from Eckel (1968):

> The Nevada Test Site represents a classic case of serendipity. Originally an Air Force bombing and gunnery range during World War II, it was converted by the Atomic Energy Commission to a proving ground for nuclear weapons in 1950. Virtually all the early nuclear explosive tests were atmospheric—devices were air-dropped or supported on steel towers or by balloons. Interest in the underlying geology of the Nevada Test Site was therefore minimal, either in original choice of the site or in early plans for its future uses. Yet when the time came for underground nuclear testing, it was soon found that the geologic and hydrologic conditions at the Nevada Test Site were ideal for the purpose. This is best indicated by the fact that many studies have been made in search of possible test sites elsewhere in the United States and its possessions. Except for a few very special purpose sites, as when great thicknesses of salt were required, not one of these studies has resulted in possible sites that quite measure up to the already existing and well-developed Nevada Test Site.

All total, 828 underground tests were made at the NTS from 1956 to 1991, when a worldwide moratorium on nuclear testing went into effect. Not all of the underground tests were weapons-related. Some were part of Project Plowshare, which involved testing for peaceful uses of nuclear bombs. One of the most noteworthy Plowshare tests was the Sedan shot (a hydrogen device) which was detonated on July 6, 1962. This shot created a crater 390 meters (1280 feet) in diameter and 98 meters (320 feet) in depth and it vented radioactive fallout over an extensive area. This shot and others like it were studied to obtain data on nuclear "earth moving" projects, like the construction of a sea-level canal across the Isthmus of Panama.

Other non-weapons work also took place at the NTS, such as the testing of nuclear reactors for space exploration.

Additionally, the U.S. Army participated in training exercises associated with the nuclear testing, including mock battles that might occur in atomic warfare, and the Civil Defense Administration became involved, for example in studying the effects of detonations on model settlements that were constructed near ground zero of bomb tests, nicknamed "doom towns" (Kirk 2012).

In retrospect it seems amazing that 928 nuclear devices (mostly weapons) were tested in the desert of southern Nevada (100 atmospheric tests and 828 underground tests). The context was the very real Cold War threats that are now fading from the collective memory of the U.S. However at the time, as noted by Fehner and Gosling (2000), "the Nevada Test Site played a vital and central role in the development and maintenance of the Cold War nuclear arsenal".

Radioecology at the NTS

The desert setting of the NTS makes it distinctive among the locations of early radioecology research. The climate, as expected for a continental desert, is harsh. For example, at Yucca Flat during the study years the recorded extreme temperatures ranged from 110 degrees F (43 degrees C) to -14 degrees F (-26 degrees C). The average annual precipitation was about 4 inches (102 mm) with measurable precipitation occurring, on average, less than 30 days per year. Furthermore, the growing conditions varied widely from year to year (O'Farrell and Emery 1976). The vegetation was dominated by low, widely spaced shrubs growing on stony soils or desert pavement (Allred et al. 1963). Typically, vegetation covered on average about 14% of the ground surface (Rickard 1963). The desert at the NTS was (and is) a tough place to do fieldwork, and ecological research was made even more difficult by the frequent bomb testing, which complicated experimental designs and limited access to field sites.

Environmental monitoring of the bomb testing started slowly at the NTS. It doesn't seem to have been a priority, perhaps for several reasons. In the 1950s relatively few ecologists were studying deserts and the ecosystem concept had not been applied in this biome. The initial work on environmental monitoring at the NTS was done primarily by scientists from New Mexico Highlands University and from Brigham Young University in Utah, but by the 1960s scientists from the newly established Radiation Research Laboratory at UCLA began to lead the monitoring studies. Most of the early monitoring research was written as unpublished reports to the AEC, some of which was reviewed by Martin (1964). Ultimately, by the 1960s, despite obstacles and restrictions, much interesting Radioecology research was done at the NTS.

Fallout

Early ecological work at the NTS focused on effects of radioactive fallout on the biota. This was logical, since the frequent bomb tests were releasing fission radionuclides that spread downwind throughout the desert. Fallout was greatest from the early atmospheric testing, but radionuclides were also released during at least some of the underground tests when venting occurred. The early research showed that retention of radioactive dust particles from fallout by the shrubs and herbaceous annuals was significant (Romney et al. 1963). This pathway of contamination was experimentally demonstrated when clear plastic covers were placed over plants in the field before detonations to keep off fallout debris (Wallace and Romney 1972a, pp. 58–64). Comparisons showed that plants directly exposed to fallout were significantly more radioactive than adjacent plants that had been covered before the detonations. One of the common adaptations of desert shrubs is the abundance of fine hairs on the leaf surfaces (Ehleringer et al. 1976, Thomson et al. 1979), which function in heat dissipation and perhaps in deterring herbivores. It was suggested that this adaptation probably enhanced retention of fallout debris to a greater extent than if this adaptation was less common (Romney et al. 1963). Experimental tests in which plants were grown in soils previously contaminated by radioactive fallout also showed that root uptake of radionuclides did occur, but the uptake rates were relatively low (Mills and Shields 1961, Romney et al. 1970).

Both experimental studies in the lab (French et al. 1965) and field studies in habitats contaminated by fallout (Turner et al. 1966, Martin 1963a, 1963b, 1965, Wallace and Romney 1972a, pp. 58–64) demonstrated that desert herbivores, especially kangaroo rats (*Dipoldomys* spp.) and jackrabbits (*Lepus* spp.), could take up the fallout radionuclides and move them along the food chain. The animals had radioactivity in bone, muscle and the gastrointestinal tract but the levels were relatively low. For example, Wallace and Romney (1972a, p. 60) state, "At no time have we detected gross anatomic lesions while processing animal samples", and Turner et al. (1966) state, "There is no evidence that jackrabbits from areas contaminated by the Sedan test were dangerously radioactive". Martin (1964, 1965) synthesized much of the early work and provided a generalized ecosystem diagram showing principal pathways of fallout materials.

There was special interest in the isotopes of iodine, which are released in large amounts into fallout by the detonation of fission devices. These isotopes are not formed immediately but result from the radioactive decay of other elements produced by the fission process. As described by French and Larson (1963):

> Iodine-131 results from the decay of at least two important precursor elements. Antimony-131 decays with a half-life of 23 minutes to tellurium-131, which in turn decays with a half-life of 25 minutes to iodine-131. This is a simplification of the actual chain, and it illustrates the delay in formation of this isotope of iodine. As a result, several hours elapse before the maximum amount of iodine-131 is available. This delay may be important to deposition in the environment."

Iodine gas or vapor is absorbed on dust particles in the atmosphere and, even though the half-life of the isotopes is relatively short (2 hours to 8 days), it is a major component of radioactive fallout. Iodine isotopes are of special interest because once the element is assimilated by the vertebrate organism, through either inhalation or ingestion, it is concentrated in the thyroid gland. I-131 is of special concern because it has the longest half-life of all of the radioactive isotopes of iodine. Damage to the thyroid is from ionization by both particles that are emitted during disintegration of the radioiodine. The human public health consequences of radioiodine in fallout from nuclear testing had been understood since early in the Cold War (Comar et al. 1957) and therefore the topic received extra attention as part of the radioecology research at the NTS. The food chain transport of fallout I-131 to grass to cows to milk and to humans was well understood (Bergstrom 1967) and, in fact, this contamination pathway was being monitored in areas receiving fallout from the NTS. For example, Figure 4.1, from Pendleton et al. (1963), depicts data for the pulse of I-131 in milk on Utah farms after the Sedan test in 1962. The data show a rapid increase after the Sedan test followed by a longer-term slow decline due to radioactive decay and other loss process. Laboratory experiments were conducted and mathematical models were constructed to understand and predict the movement of I-131 from plants to the herbivores in the desert at the NTS (Turner and Jennrich 1967, Turner 1963, 1965, French and Larson 1963, French 1967). In addition to radioiodine, occurrence and movement of other radionuclides were studied at the NTS, including plutonium (Romney et al. 1970), tritium (Koranda 1969) and strontium (Neel and Larson 1963, Martin and Turner 1966).

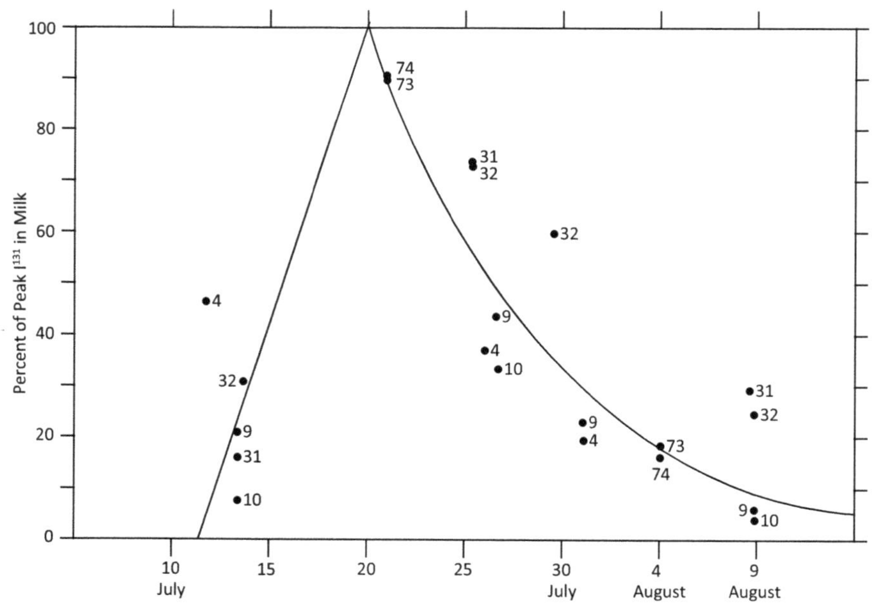

FIGURE 4.1 I-131 in milk from Utah farms during Summer 1962 originating from fallout of nuclear testing at the Nevada Test Site (from Pendleton et al. 1963).

Source: Used with permission of the AAAS.

Revegetation

The nuclear detonations, especially those done aboveground, made the radioecology experience at the NTS different than at the other sites where this kind of research was carried out through the 1950s and 1960s. A nuclear detonation causes air blast, ground shock, thermal radiation and ionizing radiation that spread out from ground zero. These phenomena had been studied at the Pacific Proving Grounds in the Marshall Islands, but there they occurred either on small islands or in the ocean. At the NTS the detonations were done in a continental setting so the full spatial extent of the bomb effects were clear to see on the landscape. A number of interesting studies were carried out at the NTS on the effects of the bomb detonations on vegetation and on the patterns of revegetation that took place afterwards. These studies are records of how the desert ecosystem self-organizes in response to bomb testing. The detonations were a new and novel disturbance for which the desert species had no previous adaptation, so the patterns of revegetation demonstrate the self-organization process that all ecosystems are capable of. This research was complicated because so many bombs were detonated in a relatively small area, most at Yucca Flat. The effects of the detonations overlapped in time and space, which eliminated the precision a scientist usually strives for in an experiment. However, over about a decade of research starting in the late 1950s Lora Shields and Philip Wells from New Mexico Highlands University along with William Rickard of the Hanford Laboratory outlined the initial response of the desert vegetation to the bomb testing. None of these early studies were sufficient to show long-term patterns, but the research has continued up to the present.

The field studies after bomb testing revealed a gradient of effects extending outward from ground zero in a general zonation pattern (Figure 4.2). In zone I there was complete destruction of all vegetation within a 0.16 kilometer (0.1 mile) radius from the target center due to severe heat, blast waves, ionizing radiation and soil displacement. In zone II, from a radius of 0.32–1.0 kilometers (0.2–0.6 miles) there was destruction of shrubs, but perennial bunchgrasses, such as *Dryzopsis hymendidies* and *Stipa speciosa*, survived, even though their aboveground crowns were burned off. In zone III, from a radius of 1.0–1.6 kilometers (0.6–1.0 mile), there was vegetation damage, but some shrubs, those with flexible stems such as *Larrea divaricate*, survived, while those with brittle wood, such as *Grayia spinosa*, were killed. No damage was evident beyond 1.6 kilometers (1 mile) from ground zero.

To study recovery, vegetation was surveyed at intervals after detonations at different test sites (Shields and Rickard 1961, Shields and Wells 1963). Often comparisons were made between spots that had been bombed and nearby undisturbed spots. The published papers on revegetation contain many good photographs of the desert vegetation with and without disturbance, which allow the reader to visualize bomb impacts. Annual plant species were the first to recolonize the middle of the bomb sites, while the normal, dominant shrub species grew inward from the outer edges more slowly. Romney et al. (1971) suggested that decades may be required for the shrub vegetation to return to its original state.

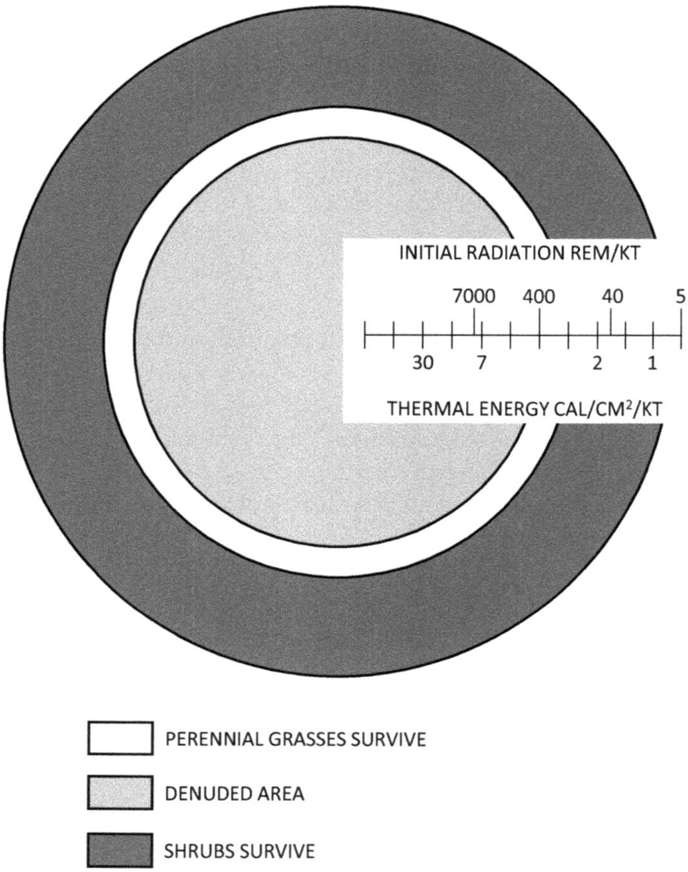

FIGURE 4.2 Zonation of vegetation around ground zero of a nuclear test at the NTS. Inset data show approximate thermal and ionizing radiation yields after the detonation (from Shields et al. 1963).

Source: Used with permission of John Wiley & Sons.

The exotic tumbleweeds, *Salsola* spp., were often the first to recolonize because "the tumbleweed habit . . . gives this species initial advantage in seeding denuded target areas" (Shields et al. 1963). Another exotic annual, *Bromus rubens*, also did well in the bombed areas. Some species, such as the native annual *Mentzeliz albica*, were found in higher abundance in the bombed areas than in the undisturbed plots, indicating that the ecosystem contained species with adaptations that allowed fast revegetation after disturbance. Beatley (1966) suggested an interesting hydrologic explanation for success of annual species based on her study of revegetation after the underground Sedan test:

> The greater survival, coverage, height and biomass of the Sedan populations during this first season following the detonation are at least in part inferred

> to reflect greater moisture availability on these sites. Here, as on most sites of surface soil disturbance in the Test Site region, infiltration of precipitation through the surface into the subsoil is apparently greater than on undisturbed sites, where a desert pavement is more or less developed and run-off is evidently greater; there is therefore greater precipitation effectiveness and consequent increased size of plants on sites of soil disturbance. In addition, where many of the shrub root systems are dead or physiologically inactive, as was the case on all three Sedan plots in the spring of 1964, soil moisture available to actively-growing plants is further increased.

However, the annual species required substantial moisture and, thus, they were susceptible to drought periods, which were common occurrences in the desert rainfall patterns.

In addition to observations made after bomb tests, Rickard and Shields (1963) also removed the surface soil in an undisturbed portion of the desert mechanically, with a bulldozer, in 1958 to mimic one of the kinds of disturbance that occurs in a bomb detonation. They followed revegetation of this "bladed" area and found that the colonizing plant species were similar to those found recolonizing the bombed area. Finally, Wells (1961) studied revegetation patterns in a ghost town, located east of Jackass Flat, that had grown up and quickly collapsed near a silver mine 33 years before nuclear testing had begun. During the period that people occupied the town the vegetation was disturbed by buildings and roads, but it recovered naturally after the people abandoned the town. This site provided "a quasi-experimental plot" for comparison with revegetation at the bomb test site. The 30-year-old plant community around the ghost town was dominated by the bunchgrass *Stipa* and two types of local shrubs, *Ephedra* and *Lycium*, that are capable of vigorous sprouting from belowground parts when the tops have been destroyed. These species are typical of desert washes that are naturally disturbed by flooding and it was suggested that the study might provide insight into long-term patterns of vegetation recovery after bomb testing.

Stationary Gamma Irradiation

There was an inherent interest in understanding ionizing radiation effects from the bomb detonations on the desert ecosystems at the NTS. It was well known that this kind of radiation causes a variety of biological injuries. However, these effects were difficult to measure and interpret at the NTS, as noted by Shields and Wells (1962):

> It has been calculated that a typical detonation of a fission-type nuclear device releases 50 percent of its energy as shock, 35 percent as heat, and 15 percent as other forms of radiation. Detonation of a nuclear weapon with an energy yield of about 40 kilotons at a point 91 meters (300 ft) above the surface eliminates all of the above-ground parts of a desert shrub

> vegetation to a distance of about 0.1 kilometer (0.05 mi) from ground zero. This denudation may be attributed to the overwhelmingly predominant thermal and shock effects. The level of ionizing radiation in the zone of complete denudation is very high, but the question of the amount of radiation injury to the vegetation is, of course, academic where the plants are largely vaporized, burned, or destroyed mechanically.

The challenge for the NTS researchers was to devise experiments on the radiation effects where possible. Early work involved radiosensitivity studies on soil fungi and algae (Durrell and Shields 1960, Shields et al. 1961) with participation of Arnold Sparrow's laboratory at Brookhaven in New York. There is also a mention of the use of a mobile Co-60 unit that generated 10,000 r for experimental irradiation in the field, but the results were not published (Brandenburg et al. 1962). Some evaluation of the direct effects was studied for the Sedan shot in July 1962 for plants and lizards. Beatley (1966) placed dosimeters on shrubs in study plots at the periphery of the blast and shock zone and followed up on annual plants that revegetated the plots with measurements with film badges for several years. Two years after the shot, regrowth by the annuals was vigorous, but a few abnormal plants were found with "greatly reduced internodes resulting in dwarfness and 'bushiness' of the plants, and failure of the corollas and other flower parts to develop". The frequency of abnormal plants was low, at 1% or less of the sampled population. Although the results were unclear due to potentially conflicting environmental influences, Beatley (1966) speculated:

> The aberrations are tentatively inferred to be the result of ionizing radiation, by way of chromosomal and/or gene mutations induced by ionizing radiation originating from nuclear devices detonated in this Test Area, including especially the Sedan detonation. Cytogenetic and morphogenetic studies of the Yucca Flat populations of this species, however, are necessary to establish or negate the validity of this inference.

The study of lizards at the Sedan shot was even more inconclusive (Turner and Gist 1965). Lizard densities pre- and post-test were assessed along a gradient from ground zero outward to more than 1.6 kilometers (1 mile). Additionally, microdosimeters were inserted under the skin of 33 lizards before the test to evaluate exposure doses. Mortality of the lizards was 100% after the shot near ground zero, but a year later young lizards were starting to return to the test area. Two lizards with dosimeters were recovered and their exposed doses were lower than the free-air dose, probably due to shielding in burrows. No conclusions about the effects of ionizing radiation could be made because of conflicting environmental factors associated with the blast.

Towner's study of chromosome cytology of the pocket mouse (*Perognathus formosus*) was more definitive. He studied populations in an area that had received significant exposure from radioactive fallout (near Yucca Flats) with a control population from an area that had received negligible fallout (near Jackass Flats) and found essentially no chromosomal aberration in either population (Towner 1965).

In order to focus on the effects of ionizing radiation on the desert, a major study was carried out in an area away from the detonations: Rock Valley, which leads to the southwest towards Death Valley, California. Four circular, 8-hectare (20-acre) study plots were established and set up for long-term studies. Three of the plots were fenced so that no terrestrial vertebrates could enter or leave the enclosures. In one of the fenced plots a centrally located Cs-137 source was installed (about 33,000 Ci) at the top of a 15-meter (50-ft) tower to generate low levels of gamma radiation. A lead shield was placed beneath the source to create a uniform field of radiation intensity throughout the plot. The free-air dose ranged from about 10 R/day near the center to about 2 R/day at the periphery of the plot, 160 meters (526 feet) away. The plot was irradiated for roughly 300 days per year, with interruptions when the source was covered so that researchers could enter the study area for measurements and sampling. Three other 8-hectare (20-acre) plots were established nearby as controls for comparison with the irradiated plot. Two of the control plots were fenced, like the radiation plot, while the third control plot was unfenced. Studies at the Rock Valley experimental site began in early 1964 and they continued for five years.

Desert shrubs in the irradiated plot were studied in relation to nuclear volume measurements that had been made before the study began (Martin 1972). Over the years of the study, effects on the shrubs were "weak and often difficult to document", as later summarized by Rundel and Gibson (1996). In 1969, after five years of exposure, three common species, those with the largest nuclear volumes, were quantitatively evaluated: *Ephedra nevadensis*, *Lycium andersonii* and *Larrea divaricata*. The evaluation was done by comparing populations in the irradiated plot with populations in one of the control plots. Only *E. nevadensis* showed significant impact from the cumulative dose of radiation, with reduced growth and fewer reproductive structures (Wallace and Romney 1972a, pp. 5–11, Kaaz et al. 1971).

In terms of animals, research focused on heteromyid rodents and lizards, which were considered to be the most important vertebrates in the study area. Three species of lizards were surveyed, but emphasis was on the most abundant species, the iguanid *Uta stansburiana*, which feeds primarily on invertebrates (mostly insects). Four species of rodents were surveyed, but emphasis was again on the most abundant species, *Peromyscus maniculatus*, which is primarily herbivorous, feeding mostly on seeds. Both of these groups of animals are burrowing and they spend significant periods of inactivity in burrows due to the harshness of the desert environment. Studies focused on population dynamics of individual species as opposed to measurements of biomass or flows of energy or nutrients.

Early work on lizards involved measuring radiation doses. Small dosimeters were placed beneath the skin of the lizards, as had been done in the study of the Sedan shot

(Turner and Gist 1965). In the field plots results showed that the lizards had less radiation exposure than expected, 30–70% less than the free-air dose, due to shielding while the animals were in their burrows (Turner and Lannom 1968). A study was also done in the laboratory of acute gamma irradiation, which showed that lethal doses were much higher than the exposures of chronic irradiation in the field (Turner et al. 1967). Initial studies in the field plots found no clear differences in lizards between the irradiated plot and the control plots for growth (Turner et al. 1965) or population parameters (Turner et al. 1969). However, as the study continued, a high incidence of female sterility was found for several of the lizard species.

Demography was emphasized in the study of radiation effects on the rodents. The dose received was assessed with small dosimeters that were attached to the skin on the backs of individual animals. The exposure dose depended on activity patterns and it averaged less than 0.1 r/day in the winter and to more than 1.5 r/day in the summer (French et al. 1966, 1974). Populations in the irradiated plot versus the control plots suggested that irradiation shortened the life span of the rodents. Interestingly, the life span in the unfenced control plot was the lowest and it was suggested that this may have been due to dispersal and predation by snakes and the kit fox, which were factors that didn't affect the populations in the fenced plots (French et al. 1967). These field data were supported by laboratory studies where the life shortening was reported to be 22–24 days of life lost per r/day of exposure (French 1969). Some of the most interesting research with the rodents involved the assessment of the intrinsic rate of increase in the population. This parameter is noted by the variable r, which is the sum of the positive instantaneous rate of births (b) and the negative instantaneous rate of deaths (d) of the population such that r = b—d. French and his collaborators used r as an index of the effect of radiation on populations. For example, French (1969) stated: "The intrinsic rate of natural increase is a measure of the success of a species in a particular environment. In this experiment a change in the rate represents a summation of the total radiation damage to the population."

The intrinsic rate of increase for the irradiated population was lower than for the control population both in the field and in the laboratory experiments, due to a lower b (birth rate) and a higher d (death rate) (French and Kaaz 1968, French et al. 1974). Despite these findings, compensating mechanisms for maintaining population size in response to low-level radiation exposure were found (French no date). These included a shift in the stable age distribution toward a greater portion of the population in more fertile age groups and increased ovulations, implantations and living fetuses.

Heritage

Radioecology research basically ended with the conclusion of the cesium source irradiation project in 1969. There were a few follow-up studies into the 1970s, but primary work was finished when the source was sealed and removed from the Rock Valley Study Site. Nuclear testing continued at the NTS until the complete global moratorium in 1991 without the level of radioecology research that took place there through the late 1950s and the 1960s. The NTS is currently operated by the Department of Energy along with the Department of Defense and it has been renamed the Nevada National Security Site. Work there includes "hazardous waste storage, conventional explosives tests, nuclear propulsion systems development, and research into remediation technologies" (Coolidge 1996).

From a public perspective the memories of the nuclear testing period at the NTS are mostly negative (Mackedon 2010, Kuletz 1998). Accidents such as fallout from the Braneberry shot (Johns and Johns 2017) and the ill-conceived atomic warfare maneuvers (Rosenberg 1980) have led to government cover-ups and a series of lawsuits. The literature on the negative effects of nuclear testing covers a broad range of reporting. At one extreme Gallagher (1993) tells intensely personal oral histories of three groups of people directly affected by the nuclear testing: people living close to the NTS (the "downwinders"), workers at the site and soldiers who were exposed to bomb testing radiation. At another extreme is the clinical records of cancers caused by fallout from nuclear testing reported by the National Cancer Institute (Institute of Medicine and NRC 1999). Native American rights associated with the NTS and the nuclear testing period were ignored, but they are now being included in considerations of the site operations (Stoffle et al. 2001). In Las Vegas, the National Atomic Testing Museum of the Smithsonian Institution, which opened in 2005, attempts to balance the negative aspects of the NTS with the positive story of its role in helping to maintain national security during the Cold War, mentioned in the introduction to this chapter. Planning for the long-term storage of nuclear wastes at nearby Yucca Mountain promises to help keep the public's attention on the heritage of nuclear testing at the NTS.

In terms of the heritage of ecological research, the radioecology program at the NTS led directly to research supported as part of the International Biological Program (IBP) at the Rock Valley Study Site. The desert was one of the five biome programs of the IBP, along with the grassland, eastern deciduous forest, coniferous forest and tundra (Norton 1974). Within the desert biome program, one main study site was chosen for each of the major desert regions in North America, and Rock Valley was selected to represent the Mojave. From 1971 to 1976 the National Science Foundation provided major funding to supplement AEC funding for fieldwork on desert ecology with associated modeling. The transition from Radioecology to the IBP Desert Biome was easily done by dropping radiation studies. Frederick Turner from UCLA, who had been a principal investigator on the Radioecology of lizards, became the IBP Program Coordinator. Research continued on vegetation, small mammals and lizards, thereby extending databases that had started with AEC funding (see, for example, Turner et al. 1982). The NTS had been termed a "natural outdoor laboratory" by Wallace and Romney (1972b) and it became "a model for long-term studies in the Mojave Desert" (Webb et al. 2009) with studies that continue to the present as part of routine field monitoring programs. One recent project has measured the effects of elevated CO_2 levels on the desert vegetation as part of global climate change research (Newingham et al. 2013).

The ecosystem concept was not an important framework for the Radioecology research of the late 1950s and the 1960s at the NTS. Most of this research involved studies of individual plant and animal species in relation to fallout, bomb detonations and radiation exposure. There was a general lack of synthesis of this work, for which the ecosystem concept could have provided a unifying theme. To some extent this issue was because the formal ecosystem concept was new and was just being articulated and made operational by the Odum brothers and others. However it may also have been caused by challenges that the desert itself makes to the ecosystem concept. Because the vegetation doesn't completely cover the ground surface and because of the relatively wide spacing and frequent periods of inactivity of individual organisms, the desert doesn't look like most other terrestrial ecosystems and, more importantly, it may not behave like them. Noy-Meir (1979/1980) felt that deserts lack the degree of integration and regulation characteristic of other ecosystems and he introduced the autecological hypothesis as an alternative description for desert life:

> In deserts, the dynamics of each population are determined mainly by its independent reaction to the environment. . . . [W]e do not need the more elaborate ecosystem hypothesis.

Whitford (2002) reviews Noy-Meir's thinking and other concepts about desert ecosystems. The IBP Desert Biome Program applied the traditional ecosystem concept broadly (Goodall et al. 1979, Goodall and Perry 1981) and made comparisons of primary production (Szarek 1979) and nutrient cycling (Wallace et al. 1978, West 1983) with synthesis that had been lacking in the older Radioecology at the NTS. More specifically, the IBP ecology research at the Rock Valley Study Site is reviewed by Rundel and Gibson (1996).

A number of taxonomic resources were developed during the radioecology research at the NTS that were useful to the IBP research and will continue to be useful for fieldwork in southern Nevada. In particular the series from Brigham Young University, based on work supported by AEC funding, is an extensive record of biodiversity at the NTS (see, for example, Gertsch and Allred 1965 on scorpions and Hayward et al. 1963 on birds). Janice Beatley, from UCLA, also made extensive studies of the vegetation at the NTS and prepared a taxonomic and natural history survey of the vascular plants (Beatley 1976). Beatley developed a sensitive understanding of the NTS vegetation as detailed in her many publications. Of special significance are the permanent plots she established (Beatley 1976), her vegetation map and extensive records of precipitation (Beatley 1974a) and her phenology model that demonstrates the role of infrequent rain events in the life histories of desert plants (Beatley 1974b). Webb et al. (2009) bring attention to the significance of Beatley's work, stating, "The work of Dr. Janice C. Beatley, a UCLA researcher, well illustrates the legacy of long-term data on the NTS and within the Mojave Desert." They continue: "Beatley's foresight has led to long-term sites of great historical value to the researchers that followed here."

In terms of the total literature, O'Farrell and Emery (1976) provide an annotated bibliography of more than 300 publications from work at the NTS, which is an impressive record of the heritage of the site. As a final note about the heritage of Radioecology at the NTS, a number of students received training there. Perhaps most significant was Clayton Gist, who started as a Brigham Young University student intern at the site and ended up working for ten years at the NTS. After that, he got his M.S. at Colorado State University and his Ph.D. at the University of Georgia in the field of Radioecology. He did field research in the El Verde project in Puerto Rico (Turner and Gist 1970), modeling for the sea level canal project (Gist 1975), worked for a time at the Kotzebue site, in Alaska, and ended up at the Oak Ridge National Laboratory, in Tennessee. Gist experienced more of the important projects in the history of Radioecology than perhaps any other individual!

Literature Cited

Allred, D. M., D. E. Beck and C. D. Jorgensen. 1963. Biotic communities of the Nevada Test Site. Brigham Young University Science Bulletin, Biological Series 2(2).

Beatley, J. C. 1966. Winter annual vegetation following a nuclear detonation in the Northern Mojave Desert (Nevada Test Site). Radiation Botany 6:69–82.

Beatley, J. C. 1969. Biomass of desert winter annual plant populations in Southern Nevada. Oikos 20:261–273.

Beatley, J. C. 1974a. Effects of rainfall and temperature on the distribution and behavior of *Larrea tridentate* (creosote-bush) in the Mojave Desert of Nevada. Ecology 55:245–261.

Beatley, J. C. 1974b. Phenological events and their environmental triggers in Mojave Desert ecosystems. Ecology 55:856–863.

Beatley, J. C. 1976. Vascular Plants of the Nevada Test Site and Central-Southern Nevada: Ecologic and Geographic Distributions. TID-26881, Energy Research and Development Administration, Washington, DC.

Bergstrom, S. O. W. 1967. Transport of fallout 131I into milk. Pp. 159–174. In: B. Åberg and F. P. Hungate (eds.). Radioecological Concentration Processes, Pergamon Press, Oxford, UK.

Brandenburg, M. K., H. L. Mills, W. H. Rickard and L. M. Shields. 1962. Effects of acute gamma radiation on growth and morphology in *Pinus monophyll* Torr. and Frem. (pinyon pine). Radiation Botany 2:251–263.

Burke, D. A. 2012. Atomic Testing in Mississippi, Project Dribble and the Quest for Nuclear Weapons Treaty Verification in the Cold War Era. Louisiana State University Press, Baton Rouge, LA.

Comar, C. L., B. F. Trum, U. S. G. Kuhn III, R. H. Wasserman, M. M. Nold and J. C. Schooley. 1957. Thyroid radioactivity after nuclear weapons tests. Science 126:16–18.

Coolidge, M. 1996. The Nevada Test Site, A Guide to America's Nuclear Proving Grounds. Center for Land Use Interpretation, Culver City, CA.

Durrell, L. W. and L. M. Shields. 1960. Fungi isolated in culture from soils of the Nevada Test Site. Mycologia 52:636–641.

Eckel, E. B. 1968. Development of geologic knowledge at Nevada Test Site. Pp. 5–19. In: E. B. Eckel (ed.). Nevada Test Site. Memoir 110. The Geological Society of America, Boulder, CO.

Ehleringer, J., O. Bjorkman and H. A. Monney. 1976. Leaf pubescence: Effects of absorptance and photosynthesis in a desert shrub. Science 192:376–377.

Fehner, T. R. and F. G. Gosling. 2000. Origins of the Nevada Test Site. DOE/MA-0518 History Division, U.S. Department of Energy, Washington, DC.

Fehner, T. R. and F. G. Gosling. 2006. Atmospheric Nuclear Weapons Testing 1951–1963. Battlefield of the Cold War, The Nevada Test Site. DOE/MA-0003 Office of History and Heritage Resources, U.S. Department of Energy, Washington, DC.

French, N. R. no date. Chronic low-level gamma irradiation of a desert ecosystem for five years. Pp. 1151–1165. In: Actes du Symposia International de Radioecologie. Volume 2. Centre dEtudes Nucleaires de Cadarache, France. Seen in: Ecological Aspects of the Nuclear Age: Selected Readings in Radiation Ecology. Compiled by V. Schultz and F. W. Whicker. TID-25978, U.S. Atomic Energy Commission, Technical Information Center, Washington, DC.

French, N. R. 1967. Comparison of radioisotope assimilation by granivorous and herbivorous mammals. Pp. 665–673. In: B. Åberg and F. P. Hungate (eds.). Radioecological Concentration Processes. Pergamon Press, Oxford, UK.

French, N. R. 1969. Radiation sensitivity of rodent species. Nature 222:1003–1004.

French, N. R., P. Hayden and T. Tagami. 1965. The source of ingested radioactivity in desert rodents. Health Physics 11:637–645.

French, N. R. and H. W. Kaaz. 1968. The intrinsic rate of natural increase of irradiated *Peromyscus* in the laboratory. Ecology 49:1172–1179.

French, N. R. and K. H. Larson. 1963. Environmental pathways of radioactive iodine from nuclear tests in arid regions. Pp. 77–81. In: V. Schultz and A. W. Klement, Jr. (eds.). Radioecology, Proceedings of the First National Symposium on Radioecology. Reinhold Publishing Corporation, New York, NY.

French, N. R., B. G. Maza and A. P. Aschwanden. 1966. Periodicity of desert rodent activity. Science 154:1194–1195.

French, N. R., B. G. Maza and A. P. Aschwanden. 1967. Life spans of *Dipodomys* and *Perognathus* in the Mojave Desert. Journal of Mammalogy 48:537–548.

French, N. R., B. G. Maza, H. O. Hill, A. P. Aschwanden and H. W. Kaaz. 1974. A population study of irradiated desert rodents. Ecological Monographs 44:45–79.

Gallagher, C. 1993. American Ground Zero: The Secret Nuclear War. Random House, New York, NY.

Gertsch, W. J. and D. M. Allred. 1965. Scorpions of the Nevada Test Site. Brigham Young University Science Bulletin, Biological Series 6(4).

Gist, C. 1975. Analog simulation of the potassium cycle. Pp. 110–116. In: F. B. Golley, J. T. McGinnis, R. G. Clements, G. I. Child and M. J. Duever (eds.). Mineral Cycling in a Tropical Moist Forest Ecosystem. University of Georgia Press, Athens, GA.

Goodall, D. W. and R. A. Perry (eds.). 1981. Arid-Land Ecosystems. Volume 2. Cambridge University Press, New York, NY.

Goodall, D. W., R. A. Perry and K. M. W. Howes (eds.). 1979. Arid-Land Ecosystems: Structure, Functioning and Management. Volume 1. Cambridge University Press, New York, NY.

Hayward, C. L., M. L. Killpack and G. L. Richards. 1963. Birds of the Nevada Test Site. Brigham Young University Science Bulletin, Biological Series 3(1).

Institute of Medicine and National Research Council. 1999. Exposure of the American People to Iodine-131 from Nevada Nuclear-Bomb Tests, Review of the National Cancer Institute Report and Public Health Implications. National Academy Press, Washington, DC.

Johns, L. C. and A. R. Johns. 2017. The Baneberry Disaster: A Generation of Atomic Fallout. University of Nevada Press, Reno, NV.
Kaaz, H. W., A. Wallace and E. M. Romney. 1971. Effect of a chronic exposure to gamma radiation on the shrub *Ephedra nevadensis* in the northern Mojave Desert. Radiation Botany 11:33–37.
Kirk. A. 2012. Rereading the nature of atomic doom towns. Environmental History 17:635–647.
Koranda, J. J. 1969. Residual tritium at Sedan Crater. Pp. 696–708. In: D. J. Nelson and F. C. Evans (eds.). Symposium on Radioecology, Proceedings of the Second National Symposium. U.S. A. E. C., CONF-670503. Office of Information Services, Washington, DC.
Kuletz, V. L. 1998. The Tainted Desert, Environmental Ruin in the American West. Routledge, New York, NY.
Mackedon, M. 2010. Bombast, Spinning Atoms in the Desert. Black Rock Institute Press, Reno, NV.
Martin, W. E. 1963a. Loss of I-131 from fallout-contaminated vegetation. Health Physics 9:1141–1148.
Martin, W. E. 1963b. Interception and retention of fallout by desert shrubs. Health Physics 11:1341–1354.
Martin, W. E. 1964. Radioecology and the study of environmental radiation. Bulleting of the Torrey Botanical Club 91:283–323.
Martin, W. E. 1965. Early food-chain kinetics of radionuclides following close-in fallout from a single nuclear detonation. Pp. 758–782. In: A. W. Klement, Jr. (ed.). Radioactive Fallout from Nuclear Weapons Tests. U.S. Atomic Energy Commission, CONF-765, Division of Technical Information. Washington, DC.
Martin, W. E. 1972. Rock Valley Project: Plans for the study of radiation effects on plant populations. Pp. 392–405. In: A. Wallace and E. M. Romney (eds.). Radioecology and Ecophysiology of Desert Plants at the Nevada Test Site. U.S. A. E. C., TID-25954, Office of Information Services. Washington, DC.
Martin, W. E. and F. B. Turner. 1966. Transfer of Sr-89 from plants to rabbits in a fallout field. Health Physics 12: 621–631.
Mills, H. L. and L. M. Shields. 1961. Root absorption of fission products by *Bromus rubens* L. from the AEC Nevada Test Site soil contaminated by an undergrounds nuclear explosion. Radiation Botany 1:84–91.
Neel, J. W. and K. H. Larson. 1963. Biological availability of strontium-90 to small native animals in fallout patterns from the Nevada Test Site. Pp. 45–49. In: V. Schultz and A. W. Klement, Jr. (eds.). Radioecology, Proceedings of the First National Symposium on Radioecology. Reinhould Publishing Corporation, New York, NY.
Newingham, B. A., C. H. Vanier, T. N. Charlet, Kiona Ogle, S. D. Smith and R. S. Nowak. 2013. No cumulative effect of 10 years of elevated [CO2] on perennial plant biomass components in the Mojave Desert. Global Change Biology 19:2168–2181.
Norton, B. E. 1974. IBP studies in the desert biome. Bulletin of the Ecological Society of America 55:6–10.
Noy-Meir, I. 1979/1980. Structure and function of desert ecosystems. Israel Journal of Botany 28:1–19.
O'Farrell, T. P. and L. A. Emery. 1976. Ecology of the Nevada Test Site: A Narrative Summary and Annotated Bibliography. NVO-167, Nevada Operations Office, U.S. Energy Research and Development Administration, Washington, DC.
Pendleton, R. C., R. D. Lloyd and C. W. Mays. 1963. Iodine-121 in Utah during July and August 1962. Science 141:640–642.
Rickard, W. H. 1963. Vegetational analyses in a creosote bush community and their radioecologic implications. Pp. 39–44. In: V. Schultz and A. W. Klement, Jr. (eds.).

Radioecology, Proceedings of the First National Symposium on Radioecology. Reinhold Publishing Corp., New York, NY.
Rickard, W. H. and L. M. Shields. 1963. An early stage in the plant recolonization of a nuclear target area. Radiation Botany 3:41–44.
Romney, E. M., R. G. Lindberg, H. A. Hawthorne, B. G. Bystrom and K. H. Larson. 1963. Contamination of plant foliage with radioactive fallout. Ecology 44:343–349.
Romney, E. M., H. M. Mork and K. H. Larson. 1970. Persistence of plutonium in soil, plants and small mammals. Health Physics 19:487–491.
Romney, E. M., A. Wallace and J. D. Childress. 1971. Revegetation problems following nuclear testing activities at the Nevada Test Site. Pp. 1015–1922. In: D. J. Nelson (ed.). Radionuclides in Ecosystems, Proceedings of the Third National Symposium on Radioecology. U.S. Atomic Energy Commission, Washington, DC.
Rosenberg, H. L. 1980. Atomic Soldiers, American Victims of Nuclear Experiments. Beacon Press, Boston, MA.
Rundel, P. W. and A. C. Gibson (eds.). 1996. Ecological Communities and Processes in a Mojave Desert Ecosystem, Rock Valley, Nevada. Cambridge University Press, Cambridge, UK.
Shields, L. M., L. W. Durrell and A. H. Sparrow. 1961. Preliminary observations on radiosensitivity of algae and fungi from soils of the Nevada Test Site. Ecology 42:440–441.
Shields, L. M. and W. H. Rickard. 1961. A preliminary evaluation of radiation effects at the Nevada Test Site. Pp. 1387–1390. In: Recent Advances in Botany. Volume 2. University of Toronto Press, Toronto, Canada.
Shields, L. M. and P. V. Wells. 1962. Effects of nuclear testing on desert vegetation. Science 135:38–40.
Shields, L. M. and P. V. Wells. 1963. Recovery of vegetation on atomic target areas at the Nevada Test Site. Pp. 307–310. In: V. Schultz and A. W. Klement, Jr. (eds.). Radioecology, Proceedings of the First National Symposium on Radioecology. Reinhold Publishing Corporation, New York, NY.
Shields, L. M., P. V. Wells and W. H. Rickard. 1963. Vegetational recovery on atomic target areas in Nevada. Ecology 44:697–705.
Stoffle, R. W., M. Nieves Zedeno and D. B. Halmo (eds.). 2001. American Indians and the Nevada Test Site. U.S. Department of Energy, Nevada Operations Office, Reno NV.
Szarek, S. R. 1979. Primary production in four North American deserts: Indices of efficiency. Journal of Arid Environments 2:187–209.
Thomson, W. W., K. Platt-Aloia and D. Koller. 1979. Ultrastructure and development of the trichomes of Larrea (Creosote Bush). Botanical Gazette 140:249–260.
Titus, A. C. 1986. Bombs in the Backyard, Atomic Testing and American Politics. University of Nevada Press, Reno, NV.
Towner, J. W. 1965. The effect of radioactive fallout at the Nevada Test Site on the chromosomes of the pocket mouse. Health Physics 11:1569–1571.
Turner, F. B. 1963. Quantitative relationships between fallout radioiodine on native vegetation and in the thyroids of herbivores. Health Physics 9:1241–1246.
Turner, F. B. 1965. Uptake of fallout radionuclides by mammals and a stochastic simulation of the process. Pp. 800–820. In: A. W. Klement, Jr. (ed.). Radioactive Fallout from Nuclear Weapons Tests. U.S.A.E.C., Division of Technical Information, Washington, DC.
Turner, F. B. and C. S. Gist. 1965. Influences of a thermonuclear cratering test on close-in populations of lizards. Ecology 46:845–852.
Turner, F. B. and C. S. Gist. 1970. Observations of lizards and tree frogs in an irradiated Puerto Rican forest. Pp. E-25–E-49. In: H. T. Odum and R. F. Pigeon (eds.). A Tropical Rain Forest: A Study of Irradiation and Ecology at El Verde, Puerto Rico. U.S.A.E.C., Office of Information Services, Washington, DC.

Turner, F. B., G. A. Hoddenbach and J. R. Lannom, Jr. 1965. Growth of lizards in natural populations exposed to gamma irradiation. Health Physics 11:1585–1593.
Turner, F. B. and R. I. Jennrich. 1967. The concentration of ^{131}I in the thyroids of herbivores and a theoretical consideration of the expected frequency distribution of thyroidal ^{131} I in a large consumer population. Pp. 175–182. In: B. Åberg and F. P. Hungate (eds.). Radioecological Concentration Processes. Pergamon Press, Oxford, UK.
Turner, F. B. and J. R. Lannom, Jr. 1968. Radiation doses sustained by lizards in a continuously irradiated natural enclosure. Ecology 49:548–551.
Turner, F. B., J. R. Lannom, Jr., H. J. Kania and B. W. Kowalewsky. 1967. Acute gamma irradiation experiments with the lizard, *Uta stansburiana*. Radiation Research 31:27–35.
Turner, F. B., P. A. Medica, K. W. Bridges and R. I. Jennrich. 1982. A population model of the lizard *Uta stansburiana* in southern Nevada. Ecological Monographs 52:243–259.
Turner, F. B., P. A. Medica, J. R. Lannom, Jr. and G. A. Hoddenbach. 1969. A demographic analysis of continuously irradiated and nonirradiated populations of the lizard *Uta stansburiana*. Radiation Research 38:349–356.
Turner, F. B., R. H. Rowland and R. A. Wood. 1966. Nuclear engineering and wildlife: Radioactivity in jackrabbits after the Sedan Test. Journal of Wildlife Management 30:433–443.
Wallace, A. and E. M. Romney (eds.). 1972a. Radioecology and Ecophysiology of Desert Plants at the Nevada Test Site. TID-25954, U.S. A. E. C., Office of Information Services, Washington, DC.
Wallace, A. and E. Romney. 1972b. Preface. P. iv. In: A. Wallace and E. M. Romney (eds.). Radioecology and Ecophysiology of Desert Plants at the Nevada Test Site. TID-25954, U.S. A. E. C., Office of Information Services, Washington, DC.
Wallace, A., E. M. Romney and R. B. Hunter. 1978. Nitrogen cycle in the northern Mojave Desert: Implications and predictions. Pp. 207–218. In: N. E. West and J. Skujins (eds.). Nitrogen in Desert Ecosystems. Dowden, Hutchinson and Ross, Inc., Stroudsburg, PA.
Webb, R. H., L. A. Defalco, T. C. Esque and P. A. Medica. 2009. A review of selected long-term ecological studies of the Mojave Desert ecosystem. Pp. 429–456. In: The Mojave Desert, Ecosystem Processes and Sustainability. University of Nevada Press, Reno, NV.
Wells, P. V. 1961. Succession in desert vegetation on streets of a Nevada ghost town. Science 134:670–671.
West, N. E. 1983. Nutrient cycling in desert ecosystems. Pp. 301–324. In: N. E. West (ed.). Temperate Deserts and Semi-deserts of the World. Volume 5. Elsevier, Amsterdam.
Whitford, W. G. 2002. Ecology of Desert Systems. Academic Press, San Diego, CA.

5
ALASKA COAST

Introduction

It was almost inevitable that atomic testing would take place in Alaska. As the Cold War continued through the 1950s, the need for testing grew and intensified. The arms race continually required the creation of new weapons that in turn required more testing. Atomic bombs came first, then long-range missiles with nuclear warheads, followed by missiles to shoot down missiles. The Pacific Proving Grounds in the Marshall Islands satisfied the initial requirements for testing but its utility was limited by logistics and sovereignty concerns. The Nevada Test Site on the edge of the Mojave Desert solved the sovereignty issue by being located in the U.S., but only low-yield weapons could be tested there due to the proximity to nearby towns and cities. Alaska solved some of these problems but it also added new ones. Ultimately all the test sites would continue to be used in some form or another even as new ones, such as Alaska, were added.

Alaska had become a state only in 1959, and at least some of its leaders saw political and economic benefits that testing might bring. Outside of the main cities of Anchorage and Fairbanks, Alaska, was sparsely populated, which generally minimized the risk of atomic testing to human safety. However, two minority groups opposed the testing and their causes would ultimately halt one of the proposed shots.

The proximity of Alaska to the Soviet Union created special geopolitical concerns. Some felt that the audacity of testing weapons so close to the Soviet Union would be a show of strength that might have positive effects. Others feared that the testing in Alaska would be seen as an act of provocation that might have negative effects. The later concern was especially relevant since the U.S. and the Soviet Union were engaged in diplomacy in an effort to create a nuclear test ban agreement.

DOI: 10.4324/9781003284901-7

An additional concern with testing in Alaska was the climate. The sites that were considered and the two finally chosen for testing were within the Arctic. The winters were long and harsh, with temperatures well below freezing. The summers, when testing could be done, were relatively mild in temperature but short. Both of these climatic extremes created special limitations on the logistic activities involved in nuclear testing. The AEC balanced all of the positive and negative factors and ended up carrying out radioecology research associated with nuclear testing in Alaska for 15 years, from the late 1950s to the early 1970s.

Project Chariot

The first radioecology project in Alaska was associated with an attempt to test the use of nuclear bombs to blast a harbor on the Chukchi Sea near the mouth of Ogotoruk Creek, located about 160 kilometers (100 miles) north of the Arctic Circle. This project was code-named Chariot and it was the first major Plowshare effort. Plowshare was the program developed by the AEC to demonstrate peaceful uses of atomic bombs. As the first Plowshare project, Chariot was considered to be a high priority. Its success was critical to AEC's goal of showing that nuclear fission and fusion technology could perform positive, non-military objectives. Plowshare started in 1957, stimulated by President Eisenhower's "Atoms for Peace" initiative. From the start the planners envisioned a sea-level canal across the Isthmus of Panama excavated by the detonation of a number of nuclear bombs as an ultimate goal. Much research was needed to understand how to carry out these explosions in a controlled fashion, and Chariot represented a first step at understanding how to use underground explosions with high-yield devices, such as the H-bomb. The work plan for blasting a harbor in Alaska was championed by Edward Teller of the Lawrence Livermore Laboratory in California. Teller was a towering figure in the field of atomic physics with a reputation as a maverick within the Manhattan Project, which helped to end World War II (Herken 2002). While everyone else was collaborating on developing a fission bomb, Teller insisted on working on a design for a fusion or hydrogen bomb, which was only a theoretical concept at the time. Now, more than a decade after fission bombs had been designed, built and dropped on Japan, Teller had helped to develop the H-bomb, and he wanted to use it for geoengineering in the Plowshare Program.

Work at the Pacific Proving Grounds had demonstrated that bombs could generate large-scale craters, but more testing was needed in order to utilize this energy source with control and precision. This was the goal of Chariot. The story of Project Chariot is described by O'Neill (1994) and much of this section relies on O'Neill's book. Teller and his group at the Livermore Lab wanted data from an underground blast of a large device to understand what might be possible with the Plowshare concept of geoengineering. They chose a blast along the coast to create a harbor because of the positive economic benefits they felt it would bring to Alaska. The site they picked was isolated, nearly 965 km (600 miles) from the large city of Fairbanks. The argument for the project was that it would stimulate economic development in an essentially underdeveloped part of the state. Starting in 1958 officials from the Livermore Lab began visits to Alaska to publicize the project and they found support among the state politicians. A minority of Alaskans questioned the project, which was in reality quite poorly planned in the beginning. In particular, a small group of biology professors from the University of Alaska raised questions that the environmental consequences of the detonation had not been adequately considered. Basically, the Livermore group claimed that no radioactivity would be released to the environment since an underground detonation was planned, and, if there were releases, winds would carry the radioactivity out to sea. Moreover, the Livermore group had basically ignored the fact that the Inuit village of Point Hope (290 residents) was located only 50 km (32 miles) inland from ground zero of the planned detonation and four other Inuit villages were within 160 kilometers (100 miles) of the site. Initially the blast was planned for the summer of 1959, but it was clear that more research was needed on the potential impact to the environment and to the Inuits in Point Hope. Therefore, field research was proposed, which included funding for work by the University of Alaska biologists.

John Wolfe Enters Radioecology

In the late 1950s the AEC lacked in-house personnel trained in Ecology and it was unprepared to undertake an environmental assessment of potential impacts from the proposed underground detonation. However, it was clear that such a study was needed for Chariot to move forward. In what turned out to be a very progressive step, the AEC hired John Wolfe, a biology professor at Ohio State University with some background in Ecology, to coordinate environmental studies at Ogotoruk Creek and Cape Thompson. Over the 15 years Wolfe came to lead a diverse program of radioecological research in essentially all the major biomes of North America. In the short run, the Chariot shot was postponed to 1961 and environmental studies supported by the AEC began in 1959. With Wolfe and an oversight committee in charge, a broad research plan was initiated.

The plan for fieldwork at Cape Thompson covered geology, especially in regard to how the rock strata might respond to an underground detonation. Since the site was coastal, both marine and terrestrial biota were surveyed. There was special interest given to potential species that might be hunted by the local people. In this regard the diet and food resources of the Inuit people in Point Hope were covered with sociological interview techniques. Finally, fallout radionuclides from past atomic testing were measured throughout the terrestrial food chain compartments. Radioactive fallout from bomb testing was consistently less in the Arctic as compared to the Temperate Zone due to global atmospheric circulation patterns. However, as will be seen, radionuclides became concentrated by ecological processes in the tundra ecosystem.

A large number of researchers were involved in the fieldwork from the base camp that was established near the mouth of Ogotoruk Creek. Due to the isolation of Cape Thompson, everything to support the research had to come by boat or by plane, including the researchers. The University of Alaska professors who were responsible for elements of wildlife biology and Inuit diet and hunting methods were likely more comfortable with the setting than many other researchers who came from agencies and universities in the temperate zone. Most of the research was carried out in the Arctic summer when occupancy of the base camp was high, but even throughout the winter a few staff stayed at the site. The research in large part was meant to anticipate possible impacts from the Project Chariot underground detonation and from any accidental venting of radionuclides to the atmosphere that might occur.

The fieldwork spanned 1959 and 1960, after which preliminary reports were released. Controversy quickly began developing on several fronts. On the national scale the preliminary AEC report on Project Chariot was compared and contrasted with a report prepared by a group of environmentalists called the Greater St. Louis Committee on Nuclear Information (CNI) in *Science* magazine. The AEC's summary ". . . found no significant radiation hazard in the project" (Margolis 1961), while the CNI report suggested that the AEC had seriously under-estimated the risk of radioactive fallout that might be produced by the detonation (Commoner et al. 1961). In Alaska, environmental groups also became more vocal in questioning the AEC's safety considerations for the project.

Within the project some scientists suggested that potential impacts of the detonation were significant, in particular in causing radioactive fallout over local caribou grazing lands and thus polluting an important food source for the Inuit in the surrounding areas. According to O'Neill (1994), Wolfe edited these opinions out of the preliminary summary report. In fact, O'Neill relates several instances in which Wolfe discounted and even covered up opinions of project scientists that opposed AEC's position that the detonation posed little or no risk. O'Neill's negative portrayal of Wolfe's role in Project Chariot is in strong contrast to subsequent descriptions that are much more positive about Wolfe's contributions (Seymour 1976, Sprugel 1975). Ultimately the uncertainty associated with underground detonation and the growing negative public opinion caused the cancellation of Project Chariot in 1962.

The AEC was not used to the scrutiny it had been receiving in terms of environmental assessment of the proposed Alaska test and they were not able to justify the risk of potential radioactivity release. This seems to have led to renewed testing in Nevada at the NTS, where they had not been held to the same degree of accountability about nuclear testing. In this regard, the massive Sedan shot on July 6, 1962 at the NTS may have taken the place of Chariot in providing information about cratering.

The report on the environmental studies carried out at Cape Thompson came out in 1966, four years after the Chariot shot had been canceled (Wilimovsky and Wolfe 1966). It was a massive text of 1250 pages with 42 chapters, covering physical conditions of the land and the Chukchi Sea, the biota and the local people. There was no detail on the Plowshare premise of the original cratering experiment, but a section on fallout radioactivity from past atomic bomb testing was included. The book was not organized with the ecosystem concept but rather utilized the theme of "bioenvironment". The Cape Thompson book was reviewed in three major journals: *Bioscience, Ecology* and the *Journal of Animal Ecology*. Bauer (1967) focused on the Plowshare connection and wondered why the book did not address the problems that caused Project Chariot to be canceled. Ball (1967) was quite positive and suggested "the book is really an encyclopedia of information about one segment of the Arctic". Elton (1967) was a bit less enthusiastic and suggested that "ecologists cannot expect a very great amount of work in depth from an expeditionary project of this nature". The latter two reviews highlighted the strengths of the chapters on caribou, seals and seabirds.

The Lichen-Caribou-Inuit Food Chain

Perhaps the most significant finding of the Project Chariot research was the importance of the food chain of lichens to caribou in biomagnifying fallout radionuclides to the Inuit. Caribou were an important food source for Inuit during at least part of the year and radionuclides that they contained would be passed on to the people who eat them. The understanding of this phenomena was one of the critical reasons that the project was canceled. The potential for fallout from the Chariot shot in irradiating the local people was too great a risk, as judged by conservationists in Alaska and, as communicated by the CNI, judged by a broader audience in the whole United States.

The ability of lichens at the base of the food chain to effectively absorb fallout radionuclides from atomic testing was the key to making the food chain toxic from radiation. Many lichens have evolved a large surface area-to-volume ratio to absorb moisture and nutrients from the atmosphere. This morphological characteristic preadapted lichens for absorbing fallout radionuclides, such as radiocesium and radiostrontium, by direct foliar interception (Gorham 1959). Also, because lichens are long-lived, they are able to retain and effectively accumulate the radionuclides. Eight lichen species were examined during the Alaska studies and the genus *Cornicularia* was found to have the highest Cs-137 concentrations. This genus grows on windswept ridgetops that are nearly free of snow during the winter. This factor causes them to receive more direct deposition of fallout radionuclides, especially in comparison to other lichen taxa that grow in habitats that are snow covered for long periods of time. Furthermore, the lack of snow cover makes *Cornicularia* more accessible to grazing by caribou, thereby increasing the flow of radionuclides along the food chain to humans.

At the start of Chariot, it was widely known that fallout radionuclides could enter and be transferred up grazing food chains to humans (Anderson et al. 1957) Thus, studies were undertaken at Cape Thompson to gather data on food chains leading to the local people living in the area. It should be noted that parallel work was being done simultaneously in the Arctic of Scandinavia, where data was showing that Laplanders, who primarily fed on reindeer, were also receiving elevated doses of fallout radionuclides that had been absorbed by lichens (see Witkamp 1966, Hanson et al. 1967 for general reviews of the Arctic food chains).

W. C. Hanson and H. E. Palmer along with other ecologists from the Hanford Laboratory in Richland, Washington, were the primary researchers involved in studying the lichen-caribou-Inuit food chain in Alaska. As part of the initial work of the bioenvironmental survey, measurements of radionuclides in biota and the local, native human population were started in 1959 and continued for nearly 15 years with AEC support in various forms long after Chariot had been cancelled. Initial studies of tundra vegetation demonstrated that the lichens and mosses had more radioactivity than vascular plants and that cesium-137 was the most important fallout radionuclide because of its relatively long half-life (Rickard et al. 1965). This data was confirmed over time and was extended to cover the local Native population of Inuit (Palmer et al. 1963, Hanson and Palmer 1965, Hanson et al. 1966). Comparisons were made between four Inuit communities, including coastal villages, like Point Hope, but also inland villages, like Anaktuvuk Pass in the Brooks Mountain Range. A shadow-shield whole-body counter was developed at the Hanford Lab and it was transported and reconstructed at each village for the measurements. In general, inland villages had the highest levels of radioactivity because the people relied more heavily on caribou in their diet, while coastal villages had relatively lower levels of radioactivity because seafood made up a large portion of the diet along with caribou. Hanson (1967, 1968) summarized the unique ecological processes that concentrate Cs-137 in the lichen-caribou-Inuit food chain and showed data from Anaktuvuk Pass to illustrate the dynamics (Figure 5.1). The graph includes seasonal variations in Cs-137 concentrations along the food chain with an increase in concentration "by a factor

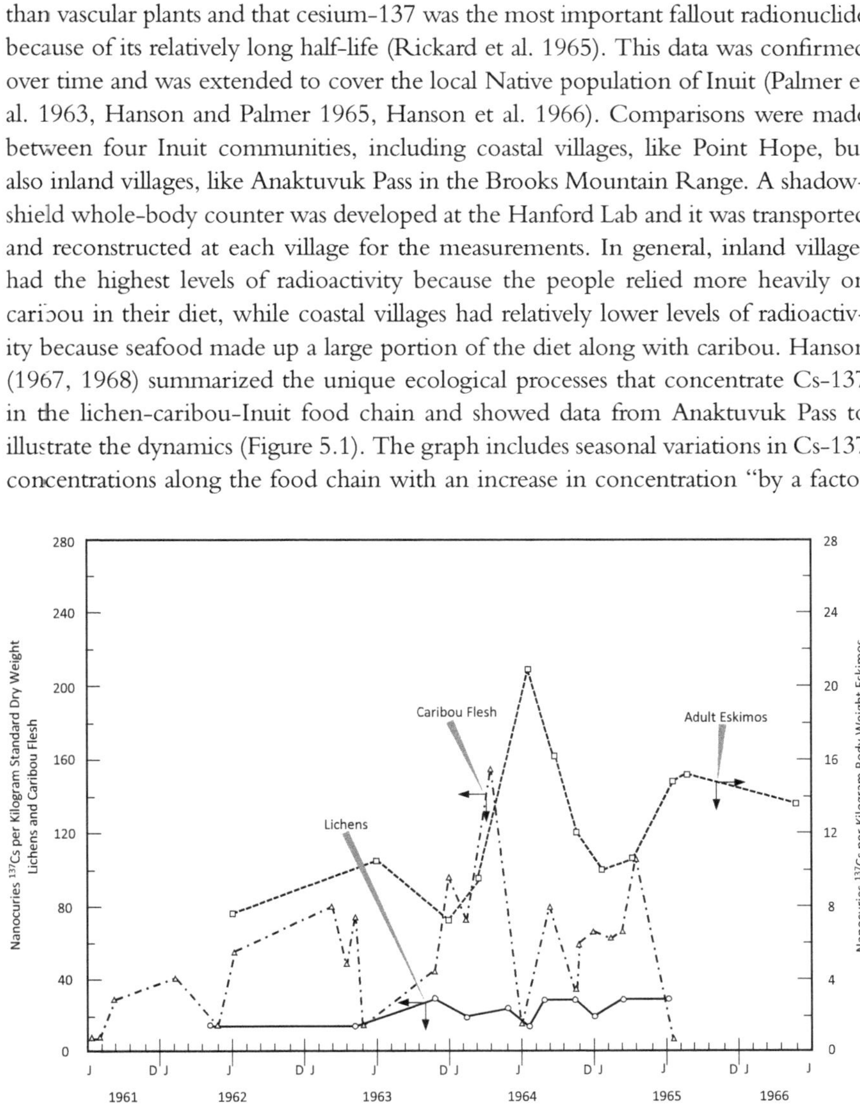

FIGURE 5.1 Cesium-137 concentrations in the lichen-caribou-Inuit food chain at Anaktuvuk Pass, Alaska, 1961–1963 (from Hanson 1967).

Source: Used with permission from Wolters Kluwer.

of about two at each successive trophic level" (Hanson 1967). Collaborating with L. Eberhardt, Hanson created a model of the food chain (Eberhardt and Hanson 1969). The model included three state variables and six transfer functions and it was used primarily for sensitivity analysis. Hansen continued to work on the Arctic ecosystems into the 1970s (Hanson 1976, Hanson et al. 1975) and received his Ph.D. degree from Colorado State University for the work (Hanson 1973).

Underground Tests on Amchitka Island

After Chariot, the next chapter in Alaskan Radioecology took place on Amchitka Island in the Aleutian Island Chain that stretches in an arc from the mainland towards the Kamchatka Peninsula of the Soviet Union. Three underground tests were done on the island in 1965, 1969 and 1971, which included two large hydrogen bomb detonations. These tests are described in Kohlhoff (2002).

Amchitka Island is one of the most southern islands in the Aleutians, located just below the Arctic Circle and about 2,160 km (1,400 miles) from Anchorage, Alaska. The island is about 65 km (40 miles) long and 7 km (4 miles) wide at the widest location, with a maritime tundra climate. It contained a naval air base with a landing strip that operated during World War II in the Aleutian Campaign when Japanese forces invaded and occupied the islands in the period 1942–1943. In addition to the landing strip, the naval air base had roads, docks, housing and a harbor, and this existing infrastructure was a factor in the choice of the island as a nuclear test site. The island had no permanent residents and the nearest non-military settlement was 450 km (270 miles) distant, at the Aleut village of Atka, with a population of 88 people (Merritt 1977). The remote location of the island was another important factor in its choice as a test site since human health risks from nuclear testing would be minimized. Additionally, a major concern, especially for the hydrogen bomb test, was the potential damage to tall buildings that could be caused by ground movements generated by the underground detonations. This risk in particular ruled out the Nevada Test Site as a location for the tests due to its proximity to the nearby city of Las Vegas, but no tall buildings were on or near Amchitka Island. The Aleutian Islands are a geologically active zone and there were continuing concerns that the underground tests might trigger earthquakes, volcanic activity or even tsunamis. The remoteness of Amchitka and the lack of a resident population helped to mitigate risks if any of these events were caused by the detonations.

A complicating factor was that Amchitka Island was part of a National Wildlife Refuge, administered by the U.S. Fish and Wildlife Service of the Department of the Interior. The refuge had been established in 1913 because of the rich fauna associated with the surrounding coastal systems, especially marine mammals and sea birds. Under the circumstances of the Cold War, the Secretary of the Interior granted permission for the tests because of their military importance. However, conservationists in Alaska and elsewhere objected due to potential environmental impacts from the tests. Of special concern was the population of sea otters (*Enhydra lutris*) living in the coastal waters around the island. Sea otters had been hunted to near extinction throughout their geographic range along the west coast of North America. A healthy population was found around Amchitka due to the efforts of the wildlife refuge staff. Therefore, this particular species became a focus of efforts to assess environmental impacts during the period of the tests.

The first test, code-named Long Shot, took place on October 29, 1965. This was an 80,000-ton blast, at the bottom of a 700 m (2,300 ft) shaft. The purpose of Long Shot was to test seismic shock sensing technologies for detecting underground explosions. There was a need to be able to differentiate underground bomb detonations from natural earthquakes in order to monitor testing. Long Shot thus had a strategic significance for the United States, which suspected that Soviet Union was cheating by holding underground tests during the International Moratorium. The timing of the Long Shot test was not ideal, since the "Great Alaskan Earthquake", the largest ever recorded in North America, had just taken place on March 27, 1964, and many people in Alaska were worried that a bomb detonation might trigger more seismic activity. Despite these concerns the test took place as scheduled without causing earthquakes and without major environmental impact. According to Kohlhoff (2002), the seismic sensing technology was not able to differentiate the detonation from a normal earthquake, but no radioactivity was released from the blast and no earthquakes were triggered, so the test was judged to be a success.

Subsequent underground detonations on Amchitka were done to test aspects of the intercontinental ballistic missile (ICBM) weapons system and its deterrent, the anti-ballistic missile (ABM). These missiles carried nuclear warheads and were developed in the arms race between the U.S. and the Soviet Union during the Cold War. By the early 1960s ICBMs had become a critical weapon in the nuclear arsenal, so much testing was required. Two underground hydrogen bomb tests were carried out on Amchitka as part of the ICBM/ABM system: code name Milrow in October 1969, a detonation of one million tons, and code name Cannikin in November 1971, a detonation of five million tons. Because of the massive explosive power of these devices, they were detonated at the bottom of shafts approximately 1,900 m (6,200 feet) deep. Shot Cannikin was the largest test ever conducted on the North American continent, second only in magnitude to Shot Mike (10 million tons of TNT force), which had obliterated Elugelab Island, on Eniwetok Atoll, in October 1951. Both Milrow and Cannikin were successful from the standpoint of the environment, because no earthquakes were triggered and, in particular, no radioactivity was apparently released in the blasts.

Environmental Assessment at Amchitka

There was much secrecy with the first detonation, Project Long Shot, and apparently little environmental assessment of the underground test. The Battelle Institute of Columbus, Ohio, was hired to carry out subsequent environmental assessments. Its work on Milrow was synthesized in a special issue of the journal *Bioscience* (Volume 21, number 12, June 1971),and later a major edited volume was prepared with information on all three shots on Amchitka (Merritt and Fuller 1977). The studies done on

Amchitka were guided by the earlier work on Cape Thompson. As noted by Giller (1977) in the foreword to the Amchitka book summary:

> There was an important precedent to such bioenvironmental investigations in Alaska in the ones that were performed as part of the AEC's project Chariot studies from 1959 to 1961. The lessons learned from Chariot were very valuable in designing and conducting the Amchitka studies.

Although Project Chariot studies apparently did influence the environmental assessment of nuclear tests on Amchitka, it is interesting to note that John Wolfe does not seem to have been involved and there is no overlap in chapter authors between the Cape Thompson book (Wilimovsky and Wolfe 1966) and the Amchitka book (Merritt and Fuller 1977).

There was no ecosystem orientation to the research plan, unlike what had been used by Battelle in its planning studies for the sea-level canal Plowshare project in Panama (see Chapter 9), but a broad bioenvironmental survey of the island was carried out. Davis (1971) described the objectives of the work as follows:

> The main objectives of the Amchitka Bioenvironmental Program are designed to:
>
> 1) Characterize the ecological features of the island and surrounding waters;
> 2) Predict, document and evaluate the effects of nuclear testing activities on the biota and the environment;
> 3) Recommend measures for minimizing adverse effects of nuclear testing activities on the biota and the environment.

Research on Amchitka began in 1967, before the Milrow shot, and concluded after the Cannakin shot, in 1973. A summary of the work is given by Merritt and Fuller (1977), with 682 pages and 26 chapters organized into six sections: The Setting, The Land, The Sea, Marine Mammals, Environmental Contaminants and Test Effects (which has only one chapter). In general it is a very good description of the ecology of the island with much background data and many diagrams of habitat cross-sections and food webs. Compartment model diagrams of terrestrial and stream ecosystems are included but do not show any radioecological interactions or effects.

Sea otters were an important part of the assessment. Two chapters about them were included in the special issue of *Bioscience* (Wright and Allton 1971, Raines et al. 1971) along with three chapters in the summary book (Estes 1977, Palmisano and Estes 1977, Morrison et al. 1977). This species is a predator in the kelp zone along the coast, eating sea urchins and other invertebrates. Figure 5.2 shows the ecological significance of sea otter predation within the coastal food web. With dense otter populations, herbivorous invertebrates are regulated

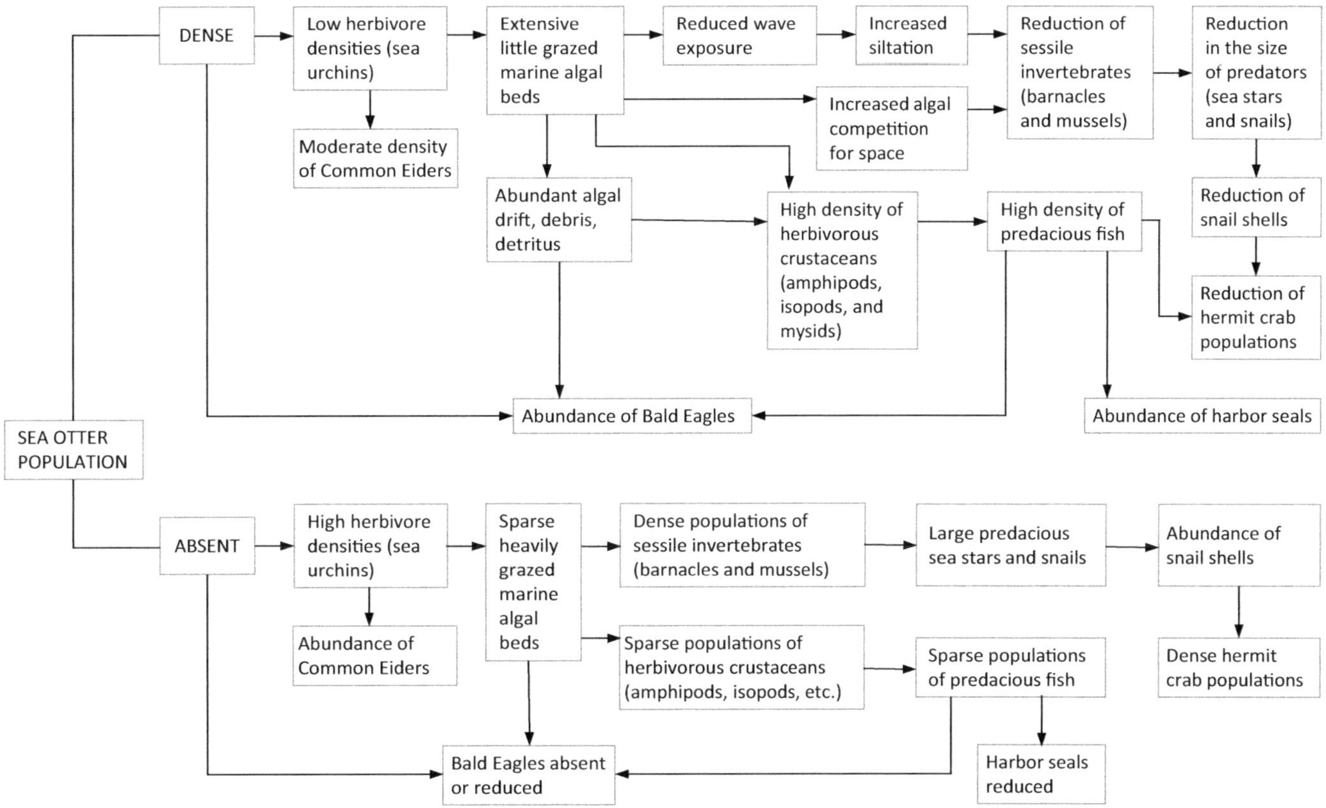

FIGURE 5.2 Causal diagram of interactions in the nearshore coastal communities with and without the presence of sea otters from (Palmisano and Estes 1977).

and many vertebrate species, such as birds and fishes, benefit from positive indirect effects, while with no sea otters the herbivorous invertebrates increase in abundance and cause negative indirect effects on marine vertebrates. The sea otter population had been decimated by the fur trade, but the otters were protected in the National Wildlife Refuge, where the populations had rebounded. As a precaution a number of sea otters were translocated from Amchitka to other sites throughout the Pacific Northwest before the underground detonation. Some otters were killed by the detonations based on pre- and post-shot population counts. For example, 21 dead or injured individuals were found after the Cannikan shot, but more were suspected to have been killed. Autopsies on the dead animals suggested they had been killed primarily by "sharp transient pressure pulses in the water with the passage of the shock wave" (Fuller and Kirkwood 1977). Overall, out of a population estimated at more than 6,000 animals it was thought that several hundred sea otters might have been killed by the Cannikin detonation, but no long-term effects were projected. This assessment was supported by a simulation model of the population that suggested it was capable of relatively fast growth rates after a disturbance, such as the bomb detonation (Raines et al. 1971).

Chapter 24 in the summary book covers the radiological assessments of the underground tests (Seymour and Nelson 1977). Samples of air, water, soils and biota were analyzed before and after the tests. The survey was extensive and much data are included in the 20 appendix tables at the end of the chapter. It was concluded that there were essentially no releases of radionuclides from the three underground nuclear detonations. The only exception was some tritium found near the Long Shot test site, though the concentrations were "not high enough to constitute a hazard to man or to native biota" (Fuller and Kirkwood 1977).

Although there was little or no direct radiological impact from the underground tests, there were other environmental impacts from associated events and activities. These impacts are described in 12 of the 26 chapters in the summary book, including the review given in Fuller and Kirkwood (1977). "Collapse craters" were formed at the ground zero locations of the detonations and shock waves caused cliff falls and turf slides, both on land and underwater, around the island. These impacts altered habitats and changed drainage patterns. Additionally, pollution was caused by effluents from the drilling of the detonation shafts. Overall, however, these impacts were not considered to be of major concern by the Battelle group (Fuller and Kirkwood 1977, Kirkwood 1971).

Of note, based on review of the political context of Amchitka tests, Kohlhoff (2002) did not agree with the conclusions of the Battelle study. He felt the "blast effects, combined with the impact of site preparation, camp construction, and drilling operations produced an ecological chamber of horrors." Furthermore, he cites correspondence from a conservationist to an Alaskan senator suggesting that the Battelle research was "a smokescreen designed to make people feel better about conditions on Amchitka". DeGroot (2005) concurred with this position, as reflected by his highly charged quote about the effects of the Cannikin shot:

> An estimated 1,000 sea otters were killed instantly, their skulls crushed by the shock wave. Other marine mammals were found with eyes blown out or lungs ruptured. Thousands of birds perished, their spines snapped or their legs pushed through their bodies from the force of the blast. Of greater long-term concern was the fact the blast ruptured the earth's crust, sucking White Alice creek into a new underground aquifer, where henceforth it became a conduit for radioactive elements. A few months later, samples taken from the blood and urine of Aleuts living nearby revealed dangerously high levels of tritium and cesium-137. Measurements taken in 1996 showed continued leakage into the Bering Sea. An unusual incidence of cancers has become apparent among the Aleuts and those who worked at the site.

Literature citations to back up these assertions are missing but the statement is consistent with Kohlhoff's description of an "ecological chamber of horrors". As often occurs in the history of radioecological projects, it is difficult for the objective reader to know what to believe.

Heritage

Radioecological research carried out in the Alaskan tundra is encapsulated in the summary texts on Project Chariot (Wilimovsky and Wolfe 1966) and on the Amchitka Island underground tests (Merritt and Fuller 1977). These books are excellent summaries of the components of two coastal tundra ecosystems, but they lack the synthesis that comes with the application of the systems concept, as had been done in several other radioecological projects of the 1960s. The research carried out at Cape Thompson and on Amchitka did little to advance the science of Ecology. No further research was carried out at the sites, and as an alternative, the International Biological Program chose to locate its major ecosystem studies on the coastal tundra at Barrow, Alaska, where fieldwork was conducted from 1970 to 1974 (Brown et al. 1980). Thus, the Alaska radioecology work performed its immediate function in describing the environments of the test sites and assessing the nuclear tests that were either proposed (Chariot) or conducted (Long Shot, Milrow and Cannakin) but it generated little in the way of heritage.

An exception is the research on the significance of sea otter predation in the kelp beds surrounding Amchitka Island. This particular research became an important case study on the concept of top-down control of food webs in the field of Ecology. The work, which originally appeared in the Battelle reports (Palmisano and Estes 1977), was also published in the prestigious journal *Science* (Estes and Palmisano 1974). The sea otters were identified as an example of a keystone species (Mills et al. 1993) in regulating the structure and function of an ecosystem, in this case the kelp beds. The sea otter work was the Ph.D. dissertation research of James Estes, and it set him on a path to become a leading expert on marine food web control, a status he has maintained for 40 years (Estes and Duggins 1994, Estes 2005, Estes et al. 2011). Stolzenburg (2008) provides a popular account of how

Estes joined the Amchitka research and conceived the sea otter research that illustrates the serendipity associated with how an individual scientist's career can unfold.

Finally, O'Neill (1994) credits the Chariot Project with a different kind of heritage. Since conservationists were able to stop the underground nuclear test at Cape Thompson, O'Neill asserts that this, at least in part, was the start of the environmental movement in the United States (as can be seen in the subtitle of his book: *H-Bomb, Inupiat Eskimos, and the Roots of the Environmental Movement*). He explicitly describes his thinking in the epilogue chapter, especially in the context of Barry Commoner's role in stopping the project. O'Neill noted, "Chariot led Commoner into environmentalism, and Commoner led others into what became known as the environmental movement." Kohlhoff (2002), furthermore, describes how the environmental organization Greenpeace originated with an effort to stop the Cannikin shot in 1971. A group of protesters in Vancouver raised money, bought a boat and tried to "sail defiantly into the waters surrounding Amchitka on the eve of the explosion." This effort was unsuccessful since the boat began to leak as they approached the Aleutians and they had to come into port for repairs, causing them to miss the detonation. Although unsuccessful, this attempted protest was the stimulus to the start of what became the global-scale organization of Greenpeace.

Literature Cited

Anderson, E. C., R. L. Schuch, W. R. Fisher and W. Langham. 1957. Radioactivity of people and foods. Science 125:1273–1278.

Ball, G. E. 1967. Arctic biology (book review). Ecology 48:323–325.

Bauer, P. S. 1967. Review of environment of the Cape Thompson Region, Alaska. Bioscience 17:120–121.

Brown, J., P. C. Miller, L. L. Tieszen and F. L. Bunnell (eds.). 1980. An Arctic Ecosystem, The Coastal Tundra at Barrow, Alaska. Dowden, Hutchinson & Ross, Inc., Stroudsburg, PA.

Commoner, B., M. W. Friedlander, E. Reiss and H. Margolis. 1961. Project Chariot. Science 134:495–496, 499–500, 502–503.

Davis, J. J. 1971. The Atomic Energy Commission's interest in Bioenvironmental Research. Bioscience 21:600–602.

DeGroot, G. J. 2005. The Bomb: A Life. Harvard University Press, Cambridge, MA.

Eberhardt, L. L. and W. C. Hanson. 1969. A simulation model of the Arctic food chain. Health Physics 17:793–806.

Elton, C. S. 1967. Review of: Environment of the Cape Thompson Region, Alaska. Journal of Animal Ecology 36:755–756.

Estes, J. A. 1977. Population estimates and feeding behavior of sea otters. Pp. 511–526. In: M. L. Merritt and R. G. Fuller (eds.). The Environment of Amchitka Island, Alaska. TID-26712 Technical Information Center, Energy Research and Development Administration, Washington, DC.

Estes, J. A. 2005. Carnivory and trophic connectivity in kelp forests. Pp. 61–81. In: J. C. Ray, K. H. Redford, R. S. Steneck and J. Berger (eds.). Large Carnivores and the Conservation of Biodiversity. Island Press, Washington, DC.

Estes, J. A. and D. O. Duggins. 1994. Sea otters and kelp forests in Alaska: Generality and variation in a community ecological paradigm. Ecological Monographs 65:75–100.

Estes, J. A. and J. F. Palmisano. 1974. Sea-otters: Their role in structuring nearshore communities. Science 185: 1058–1060.

Estes, J. A. and 23 co-authors. 2011. Trophic downgrading of Planet Earth. Science 333:301–306.

Fuller, R. G. and J. B. Kirkwood. 1977. Ecological consequences of nuclear testing. Pp. 627–650. In: M. L. Merritt and R. G. Fuller (eds.). The Environment of Amchitka Island, Alaska. TID-26712 Technical Information Center, Energy Research and Development Administration, Washington, DC.

Giller, E. B. 1977. Forward. Pp. iii–iv. In: M. L. Merritt and R. G. Fuller (eds.). The Environment of Amchitka Island, Alaska. TID-26712 Technical Information Center, Energy Research and Development Administration, Washington, DC.

Gorham, E. 1959. A comparison of lower and higher plants as accumulators of radioactive fallout. Canadian Journal of Botany 37:327–329.

Hanson, W. C. 1967. Cesium-137 in Alaskan lichens, caribou and Eskimos. Health Physics 13:383–389.

Hanson, W. C. 1968. Fallout radionuclides in northern Alaskan ecosystems. Archives Environmental Health 17:639–648.

Hanson, W. C. 1973. Fallout strontium-90 and cesium-137 in northern Alaskan ecosystems during 1959–1970. PhD Dissertation, Colorado State University, Fort Collins, CO.

Hanson, W. C. 1976. Studies of transuranic elements in Arctic ecosystems. Pp. 29–39. In: C. E. Cushing, Jr. (ed.). Radioecology and Energy Resources, Proceedings of the Fourth National Conference on Radioecology. Dowden, Hutchinson & Ross, Inc., Stroudsburg, PA.

Hanson, W. C. and H. E. Palmer. 1965. Seasonal cycle of 137Cs in some Alaskan natives and animals. Health Physics 11:1401–1406.

Hanson, W. C., H. E. Palmer and B. I. Griffin. 1966. Radioactivity in North Alaskan Eskimos and their foods, summer 1962. Pp. 1151–1164. In: N. J. Wilimovsky and J. N. Wolfe (eds.). Environment of the Cape Thompson Region, Alaska. PNE-481, Division of Technical Information. U.S. Atomic Energy Commission. Washington, DC.

Hanson, W. C., D. G. Watson and R. W. Perkins. 1967. Concentration and retention of fallout radionuclides in Alaskan Arctic ecosystems. Pp. 233–245. In: B. Åberg and F. P. Hungate (eds.). Radioecological Concentration Processes. Pergamon Press, Oxford, UK.

Hanson, W. C., F. W. Whicker and J. F. Lipscomb. 1975. Lichen forage ingestion rates of free-roaming caribou estimated with fallout Cesium-137. Pp. 71–79. In: Proceedings of the First International Reindeer and Caribou Symposium. Biological Papers of the University of Alaska, Special Report No. 1. University of Alaska, Fairbanks, AK.

Herken, G. 2002. Brotherhood of the Bomb: The Tangled Lives and Loyalties of Robert Oppenheimer, Ernest Lawrence, and Edward Teller. Henry Holt and Co., New York, NY.

Kirkwood, J. B. 1971. Bioenvironmental studies at Amchitka Island, Alaska. Bioscience 21:602–606.

Kohlhoff, D. W. 2002. Amchitka and the Bomb, Nuclear Testing in Alaska. University of Washington Press, Seattle, WA.

Margolis, H. 1961. Project Chariot: Two groups of scientists issue "objective" but conflicting reports. Science 133:2000–2001.

Merritt, M. L. 1977. Geographic setting. Pp. 1–12. In: M. L. Merritt and R. G. Fuller (eds.). The Environment of Amchitka Island, Alaska. TID-26712, Technical Information Center, Energy Research and Development Administration, Washington, DC.

Merritt, M. L. and R. G. Fuller (eds.). 1977. The Environment of Amchitka Island, Alaska. TID-26712, Technical Information Center, Energy Research and Development Administration, Washington, DC.

Mills, L. S., M. E. Soule and D. F. Doak. 1993. The keystone-species concept in ecology and conservation. Bioscience 43:219–224.

Morrison, R. R., M. Rosenmann and J. A. Estes. 1977. Sea otter metabolism and heat economy. Pp. 569–578. In: M. L. Merritt and R. G. Fuller (eds.). The Environment of Amchitka Island, Alaska. TID-26712, Technical Information Center, Energy Research and Development Administration, Washington, DC.

O'Neill, D. 1994. The Firecracker Boys, H-Bombs, Inupiat Eskimos, and the Roots of the Environmental Movement. Basic Books, New York, NY.

Palmer, H. E., W. C. Hanson, B. I. Griffin and W. C. Roesch. 1963. Cesium-137 in Alaskan Eskimos. Science 142:64–66.

Palmisano, J. F. and J. A. Estes. 1977. Ecological interactions involving the sea otter. Pp. 527–568. In: M. L. Merritt and R. G. Fuller (eds.). The Environment of Amchitka Island, Alaska. TID-26712, Technical Information Center, Energy Research and Development Administration, Washington, DC.

Raines, G. E., S. G. Bloom, P. A. McKee and J. C. Bell. 1971. Mathematical simulation of sea otter population dynamics, Amchitka Island, Alaska. Bioscience 21:686–691.

Rickard, W. H., J. J. Davis, W. C. Hanson and D. G. Watson. 1965. Gamma-emitting radionuclides in Alaskan tundra vegetation, 1959, 1960, 1961. Ecology 46:352–356.

Seymour, A. H. 1976. Dedication of symposium to Professor John N. Wolfe. Pp. xiii–xv. In: C. E. Cushing, Jr. (ed.). Radioecology and Energy Resources, Proceedings of the Fourth National Conference on Radioecology. Dowden, Hutchinson & Ross, Inc., Stroudsburg, PA.

Seymour, A. H. and V. A. Nelson. 1977. Radionuclides in air, water and biota. Pp. 579–613. In: M. L. Merritt and R. G. Fuller (eds.). The Environment of Amchitka Island, Alaska. TID-26712, Technical Information Center, Energy Research and Development Administration, Washington, DC.

Sprugel, G., Jr. 1975. Resolution of Respect: John N. Wolfe, 1910–1974. Bulletin of the Ecological Society of America 56:20–22.

Stolzenburg, W. 2008. Where the Wild Things Were, Life, Death, and Ecological Wreckage in a Land of Vanishing Predators. Bloomsbury, New York, NY.

Watson, D. C., W. C. Hanson, J. J. Davis and W. H. Rickard. 1966. Radionuclides in terrestrial and freshwater biota. Pp. 1165–1200. In: N. J. Wilimovsky and J. N. Wolfe (eds.). Environment of the Cape Thompson Region, Alaska. PNE-481, Division of Technical Information. U.S. Atomic Energy Commission, Washington, DC.

Wilimovsky, N. J. and J. N. Wolfe (eds.). 1966. Environment of the Cape Thompson Region, Alaska. Division of Technical Information PNE-481, U.S. Atomic Energy Commission, Washington, DC.

Witkamp, M. 1966. Biological concentrations of 137Cs and 90Sr in Arctic food chains. Nuclear Safety 8:58–62.

Wright, R. A. and W. H. Allton. 1971. Sea otter studies in the vicinity of Amchitka Island. Bioscience 21:673–677.

6

THE OAK RIDGE NATIONAL LABORATORY

Introduction

Oak Ridge, Tennessee, was one of the three main, original centers of work for the Manhattan Project during World War II, along with Los Alamos, New Mexico, and Hanford, Washington. It was designed and constructed primarily to provide fissionable material for atomic bombs, especially U-235. The U.S. Army Corps of Engineers, under the leadership of General Leslie Groves, chose the site because it met primary requirements of security and availability of power to run the necessary industrial processes. In terms of security, Oak Ridge was a rural landscape of farmlands and scattered small towns, located about 32 kilometers (20 miles) from Knoxville in eastern Tennessee. In terms of power requirements, Oak Ridge was located near the Tennessee Valley Authority's Norris Dam, on the Clinch River, which provided electricity from hydropower.

Land acquisition began in the fall of 1942 with the displacement of the local residents. The site was 146,000 hectares (59,000 acres) in area and it was originally named the Clinton Engineer Works, after one of the towns within the area. Three major industrial factories were planned, along with a town for resident workers. The town, which was named Oak Ridge, had to be built from the ground up, essentially. The U.S. Army Corps of Engineers attempted to design Oak Ridge to be a "typical American small town as much as possible", rather than a military base (Johnson and Jackson 1981).

The Army hired outside contractors to clear the land and to build the roads, residences, businesses and community facilities. The residents began to arrive in 1943 to work in the factories and live onsite. The population of the town grew over time and reached its peak of approximately 75,000 residents in the summer of 1945, just before the end of World War II (Johnson and Jackson 1981). The Army maintained control over the town and the residents' social activities due to

DOI: 10.4324/9781003284901-8

the need for security. Access to the site required security clearance and physical entrance to the site was regulated through gates that were guarded by a detachment of soldiers.

The primary mission of the Clinton Engineering Works was the production of U-235, which had to be separated from the more common U-238. Large-scale industrial facilities were built for two principal methods of separation: gaseous diffusion and an electromagnetic process. The gaseous diffusion plant, code named K-25, utilized a process of combining uranium gas with fluorine and forcing the mixture through perforated filters that separated the lighter U-235 from the heavier U-238 isotopes. The electromagnetic process was carried out in a factory with the code name of Y-12. This process utilized Calutrons, which were specialized cyclotrons that had been developed by E. O. Lawrence at the University of California (Calutron is short for California University Cyclotron). Ultimately, the factories at Oak Ridge produced the U-235 for the atomic bomb (code-named "Little Boy") that was dropped on Hiroshima, Japan, on August 5, 1945, with a force equivalent to 20,000 tons of TNT.

In addition to producing U-235 as a fissionable material, the Clinton Engineer Works was charged with helping to develop the technology to create plutonium, a new fissionable element. This process was carried out in an atomic pile (i.e., a graphite reactor) in which neutrons emitted from the fission of U-235 were used to convert U-238 into P-239. This work was done in the factory designated X-10, which was a model for the larger operation at the Hanford, Washington, site of the Manhattan Project. The atomic pile in X-10 was only the second that had ever been constructed, after the original one built by Enrico Fermi's team at the University of Chicago in 1939.

After the end of World War II, work at Oak Ridge was initially scaled back. Population in the town declined, though by 1950 it was still the fifth largest city in Tennessee, with 30,205 residents (Johnson and Jackson 1981). The newly formed Atomic Energy Commission took control of the facility in 1947 and renamed it the Oak Ridge National Laboratory (ORNL) in 1948. The mission of the lab changed and expanded over time, and it evolved into its present form of a "full-fledged national socio-technological institution" (Krause 1992). One of its new roles became the development of nuclear energy for peaceful purposes. For example, an important function of the lab became the production of isotopes for medical and other uses through utilization of its retooled Calutrons.

Stanley Auerbach and the Origin of Radioecology at Oak Ridge

Radioecology at Oak Ridge began in the early 1950s from concerns about possible contamination from the disposal of wastes from nuclear processing. These concerns arose from forward thinking by the Health Physics Department at the lab. The industrial-scale development of atomic energy during World War II made clear the need to protect people from exposure to unsafe levels of radiation. The field of Health Physics emerged during the Manhattan Project to meet this need (Claus 1958). Early practitioners in this field had a thorough knowledge of basic physics and were often recruited from universities where they taught and carried out research. As part of the new interdisciplinary field, the health physicists developed instruments for measuring radiation and guidelines for regulating exposure to radiation sources. Karl Z. Morgan was recruited during World War II to the Health Physics Department at Oak Ridge from an academic position and he became director of the department in 1946 (Morgan and Peterson 1999). He understood that disposal of wastes from the lab could be spreading radiation hazards into the local streams and forests, and he was instrumental in hiring a trained ecologist to study radioactive pollution. Thus, Radioecology at Oak Ridge grew from the Health Physics Department, which saw the need to look beyond human health issues and to consider possible environmental impacts of waste products from the lab.

Stanley Auerbach, a recent doctoral student, was hired in 1954 to develop an ecology program at the ORNL. Auerbach was an ideal person to hire for several reasons. He came with an excellent graduate background from the informal "Chicago School" of Ecology. He received his M.S. degree from the University of Illinois, where he worked with Victor Shelford, and his Ph.D. degree from Northwestern University, where he worked with Orlando Park. Shelford and Park were important figures, leaders in the field of Ecology during the first half of the twentieth century. Both became presidents of the Ecological Society of America (see Croker 1991 for a biography of Shelford, and note the Resolution of Respect for Park by Engelmann (1970)).

Because of his graduate background, Auerbach was a product of the latest thinking in the field of Ecology. Given the times, this meant he was well trained in what was known as autecology, basically the ecology and biogeography of single species populations, but had little exposure to the ecosystem concept, which was only beginning to emerge within the field. Auerbach's personality also made him a natural administrative leader. He led the ORNL Ecology Program from its original focus on radioecology in the 1950s and 1960s to work on environmental impact assessment, the International Biological Program and research on the global carbon cycle through the 1970s and 1980s.

The details of Auerbach's hiring were recorded by Bocking (1995, 1997) from a personal interview in 1989. One of Karl Morgan's associates, Edward Struxness, reached out to Orlando Park at Northwestern University for advice about the hire. Struxness had taken an ecology course from Park and so he had a previous connection him. Park visited the ORNL several times for discussion with the Health Physics staff and ultimately recommended Auerbach to be hired as the first trained ecologist at the ORNL. During and immediately after his graduate work, Auerbach had specialized in soil animal ecology (Auerbach 1949, 1951, Park and Auerbach 1954), which related to the concerns about radioactive pollution at the ORNL. One of the main methods of radioactive waste disposal was seepage pits (Cowser and Parker 1958) and so a need to study soil animals may have been a priority, as Howden (1957) noted: "Since one method of ground disposal of the reactor wastes may be by earthen seepage pits, ground inhabiting organisms would be among the first to be affected." The soil at several locations was obviously contaminated by radioactive wastes and, thus, Auerbach's background with soil animals may have been one of the main reasons he was hired.

The first ecological research Auerbach and his colleagues undertook at the ORNL focused on the direct toxicity of gamma radiation on soil insect populations utilizing a Co-60 source in the laboratory (Auerbach et al. 1957, Howden 1957, Howden and Auerbach 1958). This work focused on dose-response curves of survivorship, larval development and reproduction, along with descriptions of deformities caused by high doses of radiation.

During the mid-1950s Auerbach was thinking critically about the challenges of dealing with effects of radioactivity and how to use ecological research in this challenge. He seems to have arrived at a synthesis with a paper published in the Journal *Ecology* in 1958 entitled "The Soil Ecosystem and Radioactive Waste Disposal to the Ground" (Auerbach 1958). In this paper he embraces the ecosystem concept of Tansley (1935) and Evans (1956) as a unifying foundation for building the radioecology program at the ORNL. The paper includes a literature review of research that had been conducted up to that date at ORNL but, more importantly, it basically lays out a kind of plan for the future Radioecology work to be done at Oak Ridge over the next ten years. This is an early articulation of the ecosystem concept without much detail, but many of the kinds of research that would come to be important in the next decade are mentioned, including radionuclide tagging of mineral cycles, leaf litter decomposition, and energy flow in terrestrial communities. Of note, the next big phase of his group's research is anticipated: fieldwork on the contaminated bed of the drained White Oak Lake. Thus, this is an important paper in the evolution of the ecosystem concept within the history of Radioecology.

As an aside, Auerbach's reference to the "soil ecosystem" in the title of the paper is a sign of the degree to which his thinking was fresh and forward-looking. In the mid-1950s few people really knew what the ecosystem was! In retrospect, it would have been more appropriate to refer to the "terrestrial ecosystem" rather than the "soil ecosystem" since the soil is a part of the larger system. He was perhaps inspired by Fenton's (1947) use of the term "ecosystem" in regard to the soil, which Auerbach cites in his paper. However, 15 years later Martin Witkamp, one of the ecologists Auerbach hired into the ORNL radioecology group, got the terminology up to date in his review of the soil as a component of the ecosystem (Witkamp 1971a).

The White Oak Lake Bed Ecosystem

The White Oak Lake Bed was the first ecosystem that Auerbach and his team studied as a whole. It seemed to be the place where he first tried to conceptualize the ecosystem as a coherent, whole unit, as envisioned in his Ecology paper. He put together a team of scientists to do the work and they gathered information on a number of processes and components in a format that would later evolve into terrestrial ecosystem ecology (Auerbach 1959). The work done there was the first large-scale ecological field research done at Oak Ridge. However, the studies were disjointed, all done at the same location but never really integrated, relative to work that would come later in the Cesium Forest Project. The White Oak Lake Bed Project related to radioactive waste management, which was why Auerbach was hired, but the design of the research was ecological, not from Health Physics. It represented a kind of jumping-off point for what would become a highly developed model of the ecosystem of the ORNL Radioecology Group. The site of the White Oak Lake Bed is thus significant in the history of Radioecology. It was also a unique location in the spectrum of sites covered in this book where atomic energy development interacted with the landscape.

Wastewaters at the ORNL were separated into three categories, differentiated by the level of radioactive materials they contained (Morton and Struxness 1956). Wastes with the greatest radioactivity were termed high-level wastes and these were physically isolated and stored on-site. Intermediate-level wastes were discharged into constructed, open pits located in weathered shale formations, which absorbed most of the radionuclides as the waters seeped into the groundwater. The wastewaters with the least radioactivity were termed low-level wastes and these were initially disposed of by dilution through White Oak Creek and into the Clinch River (Figure 6.1). White Oak Lake was created by impounding the creek in the summer of 1943 in order to act as a settling basin for the wastes and thus to reduce the discharge of wastes to the Clinch River. The dam was located 1 kilometer (0.6 miles) above the Clinch River and the area of the impounded lake was 18 hectares

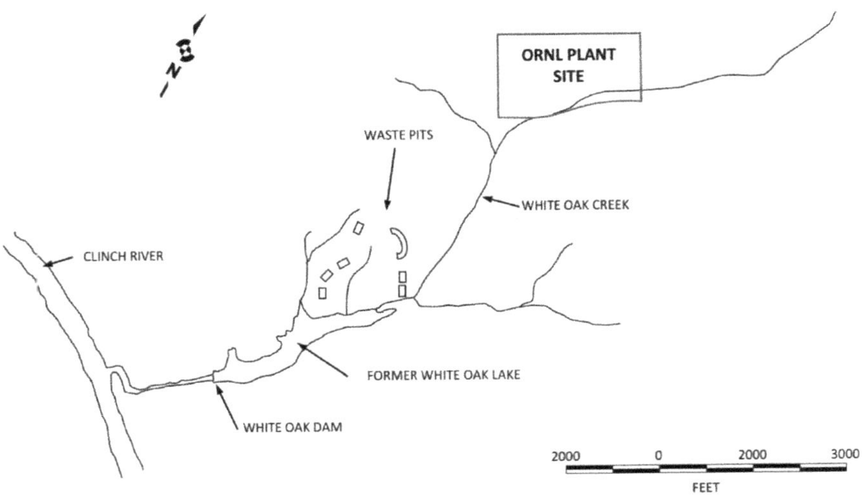

FIGURE 6.1 Location of the White Oak Lake bed at the Oak Ridge National Laboratory (from Lomenick and Gardiner 1965).

Source: Redrawn with permission of Wolters Kluwer.

(44 acres). The lake received the low-level wastes for 12 years and it, along with the Clinch River, was regularly monitored by the Health Physics Department. The lake was partially drained in Fall 1956, exposing 14 hectares (35 acres) of contaminated lake bed. Draining the lake created a new, highly contaminated terrestrial environment that offered a unique location for study. The lake bed contained large quantities of Sr-90, Cs-137, Co-60 and other isotopes that had been released as low-level wastes from the laboratory (Lomenick and Gardiner 1965). According to Reichle and Auerbach (2003),

> Even by the standards of the time, the White Oak Lake Bed was considered to be highly contaminated. . . . At the time this small piece of landscape was considered to be one of the most radioactive sites on Earth.

Up to the mid-1950s most research in radioecology had been done under controlled conditions in the laboratory, with a focus on the sensitivity of individual species to radiation. Alternatively, the drained lake bed offered an excellent location to study the effects of chronic ionizing radiation on a terrestrial ecosystem under field conditions.

The lake bed was divided into two sections based on the intensity of radioactivity. The upper end of the lake bed had higher levels of radioactivity because much of the waste loads from the laboratory settled out of the creek when it entered the impoundment. According to Kaye and Dunaway (1962), the "lower bed is not as radioactive as the upper bed which is too radioactive for personnel to work for a practical length of time." Maps of the radiation field given in the 1959 ORNL Annual Report showed the lower lake bed at 5–35 millirads/hr and the upper lake bed at 10–80 millirads/hr. Spatial variation occurred within each section, again due to differential sedimentation, with the highest activity near the stream channel and the lowest activity at the edge of what had been the lake shore when the stream was impounded.

The focus of the research on the White Oak Lake Bed in the late 1950s and early 1960s was to inventory the levels of Sr-90, Cs-137 and other isotopes in the ecosystem and to study the transfer of the isotopes along primarily the grazing food chain. A major concern was to what extent the vegetation would take up radionuclides from the soil and set in motion recycle pathways through the ecosystem and eventually to humans. This approach was tied to human health concerns, which was appropriate since Auerbach's radioecology team was a part of the Health Physics Department at the ORNL. Studies were carried out on soils, vegetation, insects and small mammals along with a single project on birds.

Natural revegetation of the lake bed took place immediately after it was exposed in 1956 (DeSelm and Shanks 1967). The vegetation was at first dominated by the forb smartweed (*Polygonum lapathipolium*), which gradually declined in coverage as a sedge-rush (*Juncus-Carex*) community and a willow (*Salix*) thicket developed. The biomass and species composition of the vegetation was monitored for five years. Plant growth was rapid due to high levels of nutrients and high moisture content of the soils. DeSelm and Shanks (1963) described the development of the willow community as follows:

> The development of the *Salix* thickets was a striking feature of vegetational change. The very dense thickets formed (140,000 stems per acre) is an exceedingly high figure. . . . The tremendous increase in mass of the Salix thicket is a phenomenon rarely reported.

Plant diversity also increased quickly: for example, from 78 taxa in 1956 to 168 kinds of woody plants in 1958 (DeSelm and Shanks 1961).

Radionuclide content of the soil and plants was sampled by several researchers with different analytical methods in an effort to understand the process of uptake and subsequent recycle of the radioactive wastes that had been deposited on the lake bed. Sr-90 and Cs-137 varied from approximately 0.1 and 0.01 uc/100 g of dry plant mass with soils at about 10 times lower activity. Graham (1958) found the physical and chemical conditions of the soil were conducive for good plant growth, with high concentrations of nitrate, sulfate, calcium and magnesium along with the radionuclides. He suggested that Sr-90 uptake by plants was inhibited by the high Ca content of the soil while Cs-137 uptake was higher because the K content of the soils was relatively low. DeSelm and Shanks (1963) multiplied concentrations of radionuclides in plant species by their biomass values and extrapolated total mass of radionuclides per unit area to be about 0.4 millicurie/hectare (1 millicurie/acre) for Sr-90, Cs-137 and Co-60 combined. Decomposition rates were not measured directly but were inferred by seasonal changes in litter mass (DeSelm and Shanks (1963). In general, the results of the early work on soils and plants on the lake bed showed that vegetation development was rapid and followed a normal succession of plant communities, despite the high radioactivity of the site. Furthermore, although there was much variability in the data, uptake and the potential for recycle of radionuclides were relatively low.

One of the main emphases in the White Oak Lake Bed Study was research on the arthropod community, especially the insects. This work was led by D. A. Crossley, who was Auerbach's first hire into the Oak Ridge radioecology group. Crossley was a recent graduate of the doctoral program in Entomology at the University of Kansas. Starting in 1956, annual surveys of arthropods were added to the studies of soils and vegetation to bring into focus the food chain transfers of the radionuclides. More than 400 species of insects were recorded and the dominant species were identified (Crossley and Howden 1961, Crossley 1963a). The taxonomic work was thorough to the degree that a new species of mite was even discovered during this work (Strandtmann and Crossley 1962). Insects were studied in the three main vegetation communities of the lake bed (sedge-rush, smartweed and willow). Total biomass of insects was estimated between 200 and 300 milligrams dry weight per square meter. The distribution of biomass and radioisotope concentration was estimated across trophic levels, including soils, plants, herbivorous and predatory insects. These surveys were first done for Cs-137 and Sr-90 (Crossley and Howden 1961, Crossley 1963a), but Ru-106 and Co-60 were added later (Crossley 1969). Most of the total mass of radionuclides in the lake bed ecosystem was held in the soils and only a small fraction of the total was found in the insects. From the initial studies Crossley and Howden (1961) concluded, "[M]ost of the radioactive materials in the lake bed soils are either not cycled through the biological system at all, or, if so, [they move] extremely slowly." Later studies (Crossley 1969) found greater "food chain mobility" of Ru-106 but because its radioactive half-life was rapid (1 year), it was not of as much concern as strontium or cesium.

More important than the static inventory of radionuclides was the measurement of their flows between compartments. Crossley modified an approach from Davis and Foster (1958) for estimating food consumption in the insect populations, based on the assertion that the body burden or whole body activity of an organism at any one time is the balance between intake and elimination. Therefore, using Cs-137 as an example, Crossley noted (1966):

> Insects feeding on radioactive plants accumulated radiocesium and reached a steady state equilibrium concentration in which intake of 137-Cs was balanced by eliminations of 137-Cs. Information on biological half-lives (eliminations rates) for cesium provided the means for estimating 137-Cs intake. Feeding rates on vegetation were calculated from Cs-137 intake rates.

This approach worked for insects because the biological half-life of Cs-137 (about 4–5 days) was much faster than the physical radioactive half-life of 33 years. The approach was confirmed with field and laboratory studies of grasshoppers (Crossley and Schnell 1961, Crossley and Pryor 1960) and with the leaf beetle *Chrysomela knabi* (Crossley 1966), which was common on the lake bed, and it became the standard for many future studies by the Oak Ridge Radioecology Group. Based on averages from fieldwork, where insects were feeding on the radioactive vegetation of the lake bed, and from carefully controlled studies in the laboratory, where insects were fed plant materials that had been artificially labeled with radionuclides, insect consumption of vegetation was estimated at 4–7% of the annual productivity (Crossley 1963b, Crossley and Howden 1961).

Small mammals were another group that was studied during the early fieldwork on the White Oak Lake Bed. This work focused on population ecology and it was led by Paul B. Dunaway and Stephen V. Kaye. Three species were studied intensively: the white-footed mouse (*Peromyscus levcopus*), which was found in the higher and drier areas were the levels of radioactivity are lower, the cotton rat (*Sigmodon hispidus*) which lives in the lower, damper areas of slightly higher radioactivity, and the rice rat (*Oryzomys palustris*), which lives in the marshy areas near the creek and is exposed to the highest levels of radioactivity (Dunaway and Kaye, 1961, 1963). The biology of the small mammals suited them to the overall objectives of the White Oak Lake Bed study. They are essentially herbivorous, which made the results of their study consistent with the grazing food chain of movements of radionuclides from soil to plants to herbivores. These animals also had limited movement patterns so their radioactivity levels would reflect the unique environment of the lake bed more closely than those of larger species that ranged more widely. Small mammal research began in December 1957 with live-trapping and mark-recapture studies. The work was carried out for three years, through 1960, when initial fieldwork ended because of a die-off caused by a severe winter. Some work was also done on different small mammal species in a nearby old field ecosystem for comparative purposes.

The overall goal of the small mammal work was to measure body burdens and to identify potential impacts of the radioactivity of the lake bed on their populations. A curious additional justification of the small mammal research was posed by Kaye and Dunaway (1963):

> Because wildlife might be utilized as an emergency source of food after a nuclear attack, ability to make predictions on the usability of this emergency source of food would be of great value.

This later justification seems a bit outlandish in the present day but, in retrospect, it reflects the nature of the Cold War era during which the research was carried out.

The fieldwork estimated population sizes, charted seasonal dynamics and recorded patterns of reproduction and body condition (Dunaway and Kaye 1964). The whole-body dose of radioactivity was estimated for cotton rats at 2.5 rads/week for external sources from the radiation field of the radionuclides in the lake bed plus an internal dose of 0.4 rads/week (mostly from Sr-90 in the bones) for a total dose rate of 2.9 rads/week (Kaye and Dunaway 1962). No statistically significant impacts of the radioactivity of the lake bed on the small mammal populations were found when comparisons with the uncontaminated old field were considered. Some differences were recorded, but they could not be attributed to radioactivity alone since other environmental conditions varied between the sites. Tables of abnormalities and lesions observed in the small mammals are given in Dunaway and Kaye (1961, 1963), but they summarized the work as follows:

> The over-all conclusion after inspection of the population data seems to be that effects of ionizing radiation on mammals from lower White Oak Lake bed and other contaminated areas are not discernible from the data.

One example of a cotton rat with skin cancer was described, but the animal was captured over 6.4 km (4 miles) from any contaminated site (Dunaway and Cosgrove 1966), so the damage may not have been caused by radioactive wastes. A possible mitigating explanation for the lack of positive findings from the radioactivity in the small mammals was that the animals' short life-span (one year) might limit the appearance of carcinogenic effects (Childs and Cosgrove 1966). Also, the studies were done in the lower lake bed, where radioactivity levels were less than in the upper lake bed area. Laboratory studies were carried out later to develop information on radioactivity effects in controlled settings and to establish radiosensitivity of the species (Dunaway et al. 1969, Kitchings et al. 1968).

A final part of the early studies of the White Oak Lake Bed was carried out by Willard (1960) on bird populations. This was a one-year study done in 1958–1959 that was supervised by E. P. Odum from the University of Georgia. The work was a cooperative project and it demonstrates Odum's connection to ORNL. Based on interviews, Stephen Bocking (1995, 1997) noted that, after he was hired, Auerbach consulted with E. P. Odum about research at Oak Ridge. The first edition of Odum's important textbook "Fundamentals of Ecology" was published in 1953 and Odum's emphasis on the ecosystem approach to research was apparently influential on the Radioecology Group at Oak Ridge. E. P. Odum was asked to serve on an advisory committee there and he is even included in a 1956 photo of the ecology group staff in front of the laboratory at Oak Ridge (Reichle and Auerbach 2003).

For his studies Willard divided the lake bed into "an inner 'bush' zone" dominated by herbaceous plants and shrubs and "an outer 'thicket' zone" dominated by young trees. Birds were collected with mist nets, counted and dissected, and radioassay was done of whole-body gross beta activity and on Sr-90 in bones and Cs-137 in muscle tissue. Species found in the inner zone, where radioactivity was higher, had higher gross beta activity. These included song sparrows (*Melospiza melodia*), field sparrows (*Spizella pusilla*), water thrushes (*Seiurus metacilla*) and chats (*Icteria virens*). In contrast, gross beta activity was lower in birds found in the outer zone, where the radioactivity was lower, including catbirds (*Dumatella carolinensis*), white-eyed vireos (*Vireo griseus*), cardinals (*Richmondona cardinalis*) and Kentucky warblers (*Oporenis formosus*). Bird species that utilized both the inner and outer habitat zones had intermediate levels of gross beta activity. Specific activity of Sr-90 and Cs-137 in birds were also found to be higher from the inner bush zone as compared with the outer thicket zone, though the overall radioactivity in the birds was very small on an area basis. However, for context, it was noted that the concentrations of Sr-90 found in the White Oak Lake Bed birds were much higher than what was considered permissible in humans.

In conclusion, the White Oak Lake Bed research was the first ecosystem-scale study carried out by the ORNL Radioecology Group. Although no formal, synthetic model came from the work, the results were in the form of a traditional ecosystem analysis: inventories of radionuclides in different ecosystem components and studies of uptake and transfers along the grazing food chain. The distribution of radioactivity of the two major radionuclides in the White Oak Lake Bed was a benchmark record of a unique ecosystem (see Table 6.1).

Research on the White Oak Lake Bed revealed an interesting example of self-organization of a natural ecosystem with high levels of radioactive contamination. There was much variability in the field data, but the ecosystem that developed on the highly radioactive site could generally be called normal, with relatively high primary productivity and diversity across a number of taxonomic groups. For example, Plummer et al. (1965) found that sedges (*Carex* spp.)

TABLE 6.1 Distribution of radioactivity from the two major radionuclides in the White Oak Lake Bed ecosystem in units of microcuries/hectares.

Component	*Sr-90*	*Cs-137*	*Reference*
Soil	600,000	12,000,000	Auerbach (1959)
Plants	3,300	550	Auerbach (1959)
Insect herbivores	0.21	0.22	Crossley and Howden (1961)
Insect predators	0.031	0.037	Crossley and Howden (1961)
Cotton rats	0.37	0.035	Kaye and Dunaway (1962)
Birds	0.017	0.047	Willard (1960)

growing on the White Oak Lake bed had longer inflorescences than sedges growing on nearby uncontaminated sites but they were otherwise normal. Also, no deformities were noted in the insect samples from the lake bed vegetation, unlike what had been found in the lab experiments that had been carried out earlier (Howden 1957, Howden and Auerbach 1958). Crossley and Howden (1961) described the system as follows:

> The resulting ecosystem on White Oak Lake Bed may be envisioned as a gigantic radioactive tracer experiment, which can yield information of interest to both ecologists and health physicists."

This characterization seems accurate and the research carried out there provided a foundation for future research by the ORNL Radioecology Group.

The Cesium Forest

While research was underway on the White Oak Lake Bed, a number of other, independent studies were begun as new personnel were added to the radioecology group. Some early work on trees related to seepage from the ponds that were holding intermediate-level wastewaters from the laboratory. Monitoring of wells around the ponds revealed that most of the radionuclides from the wastewater were being absorbed by the clay soils as it seeped into the surrounding groundwater. Only Ruthenium 106 was found moving away from the ponds in the groundwater and this led Auerbach's group to begin monitoring trees surrounding the ponds. Little was known about this radionuclide, but it was found in high concentrations in leaves, twigs and bark of trees surrounding the waste pits and in the leaf litter from these trees (Auerbach and Olson 1963). A diagram describing these findings was included in E. P. Odum's textbook as an example of early radioecology research. As noted by Reichle and Auerbach (2003):

> The positive result of this investigation is that it identified and supported the need for researchers to carry out forest experiments on the behavior and cycling of fission products, using single species of radioisotopes in

> individual tree species (Witherspoon 1963). The individual tree experiments ultimately led to the 1962 Cs-137 forest ecosystem experiment, in which an entire forest ecosystem was experimentally labeled with a radioisotope (Auerbach et al. 1964).

Work on the radioecology of Oak Ridge forests represented a new objective for the radioecology group that complemented the applied research associated with the Health Physics Department. The emphasis shifted to using radionuclides as tracers of mineral cycling pathways of the forest ecosystem. Studies continued, associated with sites that were contaminated with radioactive wastes, but new studies were started in undisturbed sites where radioisotopes were added to the ecosystem and their movements were traced to demonstrate the mineral cycles.

Witherspoon's (1963, 1964) study of the cycling of Cs-134 in white oak trees was an important step towards the new objective. Three individual oak trees on each of four sites with different soil types were experimentally inoculated with 2.0 mc of Cs-134. The work, conducted from Spring 1960 to Fall 1961, tested methods for injecting a radionuclide into a tree and for tracing its movements through the forest. Following injection into the tree trunk's cambium, Cs-134 was found to move rapidly through translocation to the leaves in the canopy and ultimately to the surrounding litter layer after leaf fall at the end of the first growing season. The radionuclide reached the soil through both leaching and leaf decomposition, and it was measured in adjacent understory vegetation in the second growing season, which documented a full cycle of the element through the forest system. This work and a number of other, separate studies on forest ecology, including leaf decomposition rates, the food web of soil animals and the application of computer simulation modeling, are described in the annual reports of the ORNL Radioecology Group. In 1962 all of these elements were combined at a single site, which came to be known as the "Cesium Forest". Previously most radioisotope tagging had been studied along food chains in which the movements of isotopes were followed from plants to herbivores to top carnivores in feeding interactions (as had been done in the White Oak Lake Bed studies and in many other spots). This kind of study of trophic transfers was important for understanding the consequences of radioactive fallout from bomb testing or of radioactive pollution from nuclear reactors. However, the research in the Cesium Forest switched the focus to transfers of radioisotopes that take place in mineral cycling of an ecosystem rather than on a one-way flow along food chains. This approach had been taken in aquatic ecosystems earlier, as in Hutchinson's study of phosphorus dynamics in Lindley Pond, but the ORNL Cesium Forest study was the first application of radioisotope cycling in a terrestrial ecosystem. The idea was to follow in detail a single radioisotope in transfers through various compartments in the forest, including the soil. All of the necessary methods had been worked out in various separate studies of the forests of the ORNL, and now they were put together to make a detailed study at a single site.

One of the unique features of the Cesium Forest was the scale of the study. Thirty-five individual trees (of which 30 were yellow poplar, *Liriodendron tulipifera*) in an area of 20 meters (66 feet) x 25 meters (82 feet) were labeled. Previously this kind of radioactive tagging had been done with one or a few individuals but the Cesium Forest study dramatically expanded the spatial scale: hence in the first publication of the work, it was referred to as a "landscape investigation" (Auerbach et al. 1964).

The site chosen for what was planned to be a long-term study was a flat-bottomed sinkhole depression in an isolated part of the ORNL reservation. This type of geomorphic setting had no surface runoff (i.e., internal drainage) and was underlain by a deep colluvial soil that was known, based on previous work, to absorb the cesium before it could reach the groundwater.

The objectives of the study were:

> . . . to determine the cycles and rates of cycling of cesium and related elements in a mesic forest ecosystem, to measure cesium transfer between compartments of an ecosystem, and apply these results to a theoretical model of element circulation developed by one of us, and to derive information useful for predicting the fate of problem radionuclides in those ecosystem which potentially may have a large capacity for storage in biological materials.
>
> *(Auerbach et al. 1964)*

All of the items in the work plan were basically new to the science of Ecology, especially the application of a theoretical model, meaning a computer model, of the mineral cycle.

Cesium was added to a trough that was constructed around the tree trunk with a sealing compound and aluminum foil to radiolabel the trees. The outer bark was smoothed in the trough and slits were chiseled through the inner bark to facilitate translocation. This method had been known for at least a decade (Graham 1954) and had been tested at ORNL in Witherspoon's (1964) white oak study. The quantity of radiocesium added to each tree was proportional to estimated tree mass based on an empirical relationship; the amount per tree ranged from 0.4 mc to 94.2 mc with a total of about 467 mc added to the entire stand (Auerbach et al. 1964). For safety, the actual procedure for adding the cesium to a tree involved four people:

> One poured the isotope, moving around the trunk to distribute the solution evenly in the trough, a second followed holding a plastic bag underneath to catch any spill while a third continued to add diluent solution at a rate equal to the quantity being absorbed by the tree. A fourth man monitored the operation with an ionization chamber ("Cutie pie") and timed it.

A photograph of the operation is shown in Figure 6.2.

The study of the movements of the cesium in the forest was led by the two new hires. Jerry Olson was hired by Auerbach in 1958, based on his experience in modeling nutrient dynamics in terrestrial ecosystems (Kwa 1993). Olson had gotten

FIGURE 6.2 Procedure for labeling a tree with Cs-137 in the Cesium Forest

Source: Photo from the Oak Ridge National Laboratory.

his doctoral degree from the University of Chicago in 1951 from work on plant succession at the Indiana Dunes, so he shared with Auerbach the background of the Chicago school of thought on Ecology. Another important addition to the Oak Ridge Radioecology Group was Martin Witkamp, from the Netherlands, who joined in 1960. He was a soil microbial ecologist and his contributions were important in the Cesium Forest research.

To track movement of cesium, the forest was heavily instrumented for sampling. Sampling of cesium in the tree canopy was done with a tower of scaffolding up to 10.7 meters (35 feet) above the ground and a hydraulic boom and basket for measurement up to 27 meters (90 feet). Sampling was also done for leaching of cesium in throughfall of rainfall and with rainfall collectors and for litterfall of leaves with litterfall traps. Finally, belowground sampling was done for soil and tree roots with forest floor sampling boxes.

Cesium was quickly translocated from the tree trunk where it was introduced up to the canopy within one month and then it was tracked in leaching and litterfall to other forest components. By the end of the first growing season about one-third of the radiocesium had transferred to the forest floor, while the remainder was reabsorbed back to the woody tissue. Storages in the tree roots were found to be important (Waller and Olson 1967) along with microbial immobilization in the litter and surface soils (Witkamp and Barzansky 1968, Witkamp 1968, Witkamp and Frank 1964).

Figure 6.3 shows the overall pattern of movements during the first growing season. Great effort was made to measure the flows and storages of cesium as it moved through

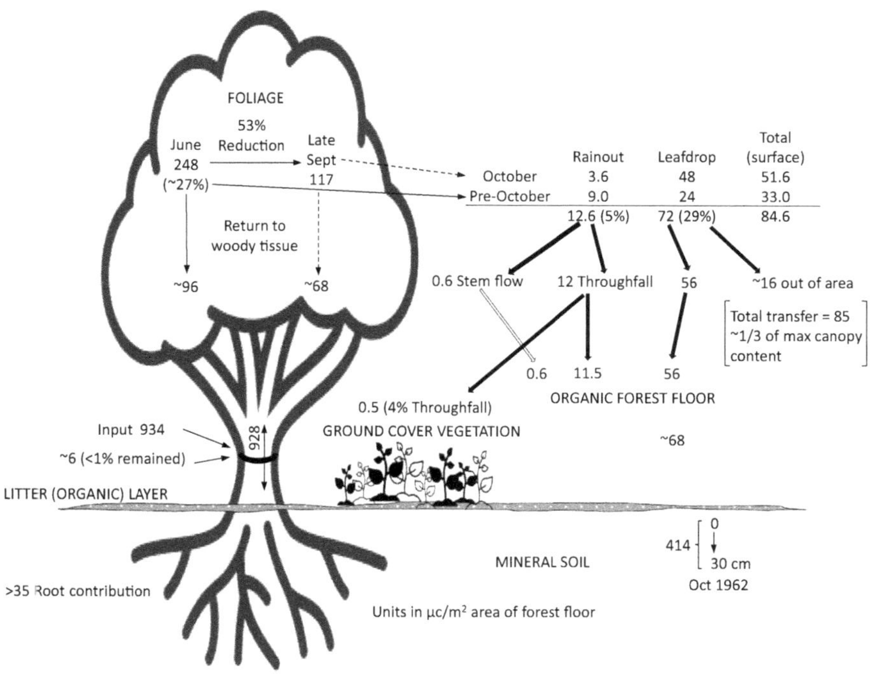

FIGURE 6.3 Model of the flows of Cs-137 in the Cesium Forest (from Waller and Olson 1967).

Source: Redrawn with permission of John Wiley & Sons.

the forest (Auerbach et al. 1967). One example of these detailed measurements is the Cs-137 found in sweet gum flower parts, given by Witherspoon and Brown (1965). The speed of movement and the diversity of compartments where Cs-137 was stored during cycling were important findings during the first few years of study.

Movement of Cs-137 from tagged leaves that fell to the forest floor at the end of the growing season to the soil through decomposition was an important transfer in the mineral cycle. A number of studies had been conducted on decomposition before the Cesium Forest study by using what became later the standard litter bag method (Bocock and Gilbert 1957). In this method a known amount of leaf material was placed in nylon mesh bags that were incubated on the forest floor. The decomposition rate of the leaf litter was measured by periodically collecting bags and recording the loss in mass and the change in chemical composition of the leaf litter. Olson and Witkamp were involved in early studies of decomposition in several forest types (Shanks and Olson 1961, Olson and Crossley 1963, Witkamp and Olson 1963, Witkamp 1966) and the role of soil animals was also studied (Crossley and Hoglund 1962), so the method was well understood. This work had been done both with leaves that had been tagged with radioisotopes and with normal leaves that had not been tagged. Leaf litter decomposition was also studied in the Cesium Forest to understand the movement of Cs-137. Witkamp (1969a, 1969b) focused on some of the factors controlling the rate of decomposition and he identified eighteen species of microbes and fungi associated with litter breakdown and measured their Cs-137 contents. He also measured the overall respiration rate of the forest floor community by CO_2 evolution, finding a diurnal cycle correlated with temperature (Witkamp 1969c).

An important focus in the Cesium Forest was the study of invertebrate animal populations, especially in the leaf litter and soil. Crossley was originally involved, but Auerbach hired David Reichle to cover this area within the ORNL Radioecology group. Like several other hires, Reichle had connections to Auerbach from his graduate degree. He had gotten his Ph.D. from Northwestern University, like Auerbach, by studying *Pselaphid* beetles (Reichle 1964) and Orlando Park had been his major professor there. Reichle embraced the radioecology techniques for studying metabolism of invertebrates developed at ORNL (Reichle 1967, 1969), and he probably benefited by initially working with Crossley, who ultimately left Oak Ridge for the University of Georgia.

Starting in Fall 1964 the cryptozoa (the animals living in the leaf litter and upper soil layers) were sampled with 40 pitfall traps that were distributed in the Cesium Forest. Cs-137 concentrations in food items (leaf litter, moss, roots, soil and wood) and in the invertebrate populations were measured to delineate the food web and to estimate rates of food consumption (Reichle and Crossley 1965, 1969). Ants and springtails were the most abundant animals found during the study, but Cs-137 concentrations were traced in at least 27 taxonomic groups. The cryptozoan food web of dominant invertebrates on the forest floor of the Cesium Forest is shown in Figure 6.4 (Reichle and Crossley 1969). The level of detail in the study of the forest floor animal populations was amazing. For example, Reichle (1968) measured oxygen consumption

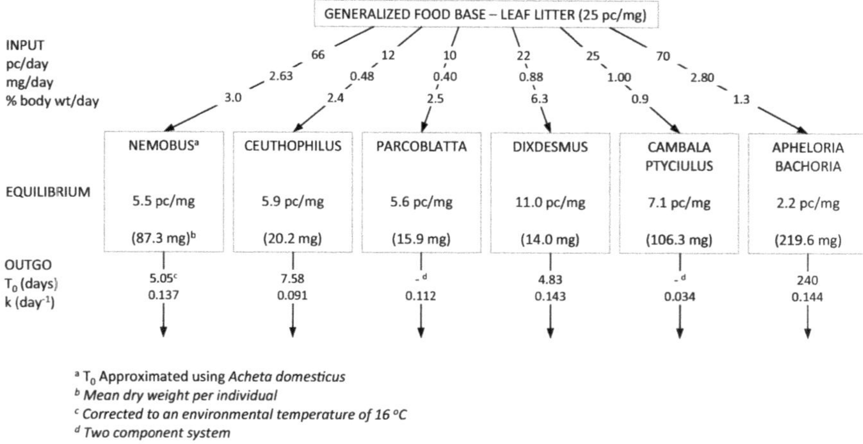

FIGURE 6.4 Leaf litter consumption by invertebrates in the Cesium Forest (from Reichle and Crossley 1965).

Source: Redrawn with permission of Wolters Kluwer.

and food intake of the small organisms that ranged in size from 2.1 milligrams to 1.6 grams live weight. Density, biomass and Cs-137 contents of forest floor invertebrates were studied over several years after the forest had been tagged in 1962, including work done by graduate students supported by ORNL on laboratory microcosms of the forest floor (McBrayer and Reichle 1971) and on the role of spider predation (Moulder and Reichle 1972). Research was also carried out on herbivorous insects in the canopy of the Cesium Forest that provided data on energy and radiocesium flow along the grazing food chain (Reichle and Crossley 1967).

One of the original stated objectives of the Cesium Forest project was to construct mathematical models of cesium cycling and to simulate them on a computer. Jerry Olson took the lead on this objective. Inspired by Hans Jenny's (1941) quantitative approach to soil development, Olson had constructed mathematical models of succession on the Indiana Dunes in his dissertation research (Olson 1958). He continued and expanded this line of research once he got to Oak Ridge; for example, he describes a five-compartment model of the flow of carbon through a forest ecosystem in the 1959 ORNL Annual Report (see also Olson 1963a, 1963b). Olson was fortunate in having access to computers at the ORNL. This kind of computing resource was not widely available to ecologists in the late 1950s and early 1960s, but it was needed for operating the nuclear technology at Oak Ridge. He first used analog computers to simulate carbon and carbon-14 in ecosystems (Olson 1963a, Neel and Olson 1962). This work represented some of the first examples of system ecology within the field of Ecology (Kwa 1993). He also helped in developing programs for simulating compartment models on digital computers (Bledsoe and Olson 1970) and he used this approach to simulate cesium cycling in the experimental forest (Olson 1965). The output of his model tracked closely what actually happened

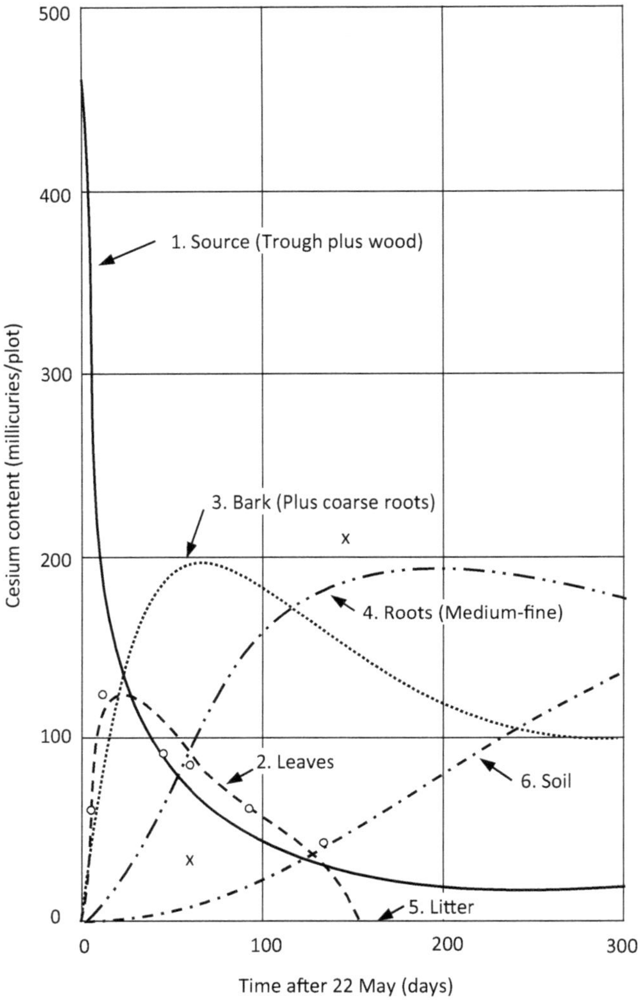

FIGURE 6.5 Analog computer simulation of Cs-137 flows through the Cesium Forest (from Olson 1965).

Source: Redrawn with permission of Wolters Kluwer.

in the main compartments of the Cesium Forest during the first year of the project (Figure 6.5). The pattern is a series of graphs that rise as cesium enters the compartment and then fall as it moves through to the next compartment, as the pulse of input moves through the forest. This modeling work helped to synthesize the separate field studies and it created a foundation that other ORNL researchers would build on in the future. Robert O'Neill, who joined the ORNL radioecology group in 1968, elaborated on this compartment model to include some of the dominant forest floor invertebrates (O'Neill 1971). The hiring of O'Neill seemed to follow a pattern that Auerbach had used before. O'Neill had done his doctoral research on forest floor millipedes (O'Neill 1967, 1968), which was close to Auerbach's own background, and he graduated from the University of Illinois, which was in effect a suburb to the Chicago school of Ecology. O'Neill's computer model of cesium movement through several forest floor invertebrates (cricket and centipede) was generalized and not matched with field data from the project, but it increased the number of compartments that were being modeled. Hoffman (1972) also extended Olson's early forest model to include cryptogams (mosses and lichens) and simulated the radiocesium distribution in the forest from 1962 to 1970. His modeling results showed that the forest floor mat and epiphytic plants on tree trunks absorbed and retained Cs-137 that moved out of the tree vegetation over time.

A fascinating set of experiments was done with microcosms of the forest floor that utilized computer modeling led by B. C. Patten and M. Witkamp. In this work microcosms with different levels of complexity were tagged with radiocesium to study the dynamics of the mineral cycle. In the first experiment five compartments of the forest floor were followed for 18 days (Patten and Witkamp 1967). The compartments were Cs-134 labeled tree leaves, mineral soil, microflora, millipedes and an aqueous leachate. Each microcosm had leaves and leachate but then other compartments were added to create a series of microcosms of increasing complexity. The combinations were:

I = leaf litter + leachate
II = leaf litter + leachate + soil
III = leaf litter + leachate + microflora
IV = leaf litter + leachate + soil + microflora
V = leaf litter + leachate + microflora + millipedes
VI = leaf litter + leachate + soil + microflora + millipedes

Triplicates of each type of microcosm were constructed and sterile water was added every 2–3 days to approximate precipitation in the Oak Ridge area. In this first experiment, oak leaves from trees inoculated during Witherspoon's study were used as input of radiocesium to the systems. After 18 days the microcosms were dissected and the content of radiocesium was measured in each of the compartments to examine the results of transfers that had occurred. Rate constants were determined by fitting models to the data with an analog computer. Each microcosm type had different rates of transfer of the radiocesium from the leaf litter to other compartments and the patterns of transfer rates were used to discuss the roles of the different compartments in the mineral cycle. For example, the microcosm type (V), with both microflora and millipedes, had the fastest loss of radiocesium from tree leaves in comparison with the other microcosm types. Networks of radiocesium transfer were identified and quantified, demonstrating the process of self-organization of the mineral cycles. In this case organization was the pattern of connections between compartments that determined the rate of mineral cycling. Various interesting analyses of the microcosm experiment were described in terms of network structure.

Additionally, the fitted equations for each microcosm type were simulated to study the dynamic behavior in terms of six quantitative attributes. The last sentence of the paper provides a forecast of the direction that the field of systems ecology would take:

> The multiplicity of material transfers and interactions conceivable in macroecosystems, together with the effects of intrasystem coupling as revealed in this investigation, make it apparent that to understand ecosystems ultimately will be to understand networks.
>
> *(Patten and Witkamp 1967)*

The initial experiment was followed by two others with changes to microcosm complexity, length of incubation time and measurements of system function (Witkamp and Frank 1969, 1970). In these experiments radiocesium was added to the microcosms in sweet gum leaves that had been tagged with Cs-137 from the Cesium Forest plot. Decomposition rate of the leaf litter was measured, along with CO_2 evolution as a measure of respiration rate of the microcosms. The results again showed self-organization of the compartments within the microcosms. These studies with microcosms of the forest floor added to the understanding of the Cesium Forest in ways not possible with field studies alone.

A number of other studies of radioecology of terrestrial ecosystems were carried out at Oak Ridge. These included research on the effects of fast neutron doses on tree species (Witherspoon 1965), dynamics of an old field contaminated by artificial fallout (Dahlman et al. 1969, Van Hook 1971) and cycling of radiocalcium by dogwood trees (Thomas 1969). One of the most unusual was the study of early stages of revegetation at nearby Copperhill, Tennessee, where sulfur dioxide from copper smelters had destroyed the surrounding forest (Witkamp et al. 1966). This particular work had no direct connection with radioecology, but the denuded site was relevant as a possible analog of a landscape devastated by impacts of nuclear bomb detonation or from radioactive wastes. This and all of the other terrestrial ecology studies demonstrate the depth and breadth of the Radioecology program at Oak Ridge through the 1960s.

Aquatic Ecology at the ORNL

Studies of aquatic ecosystems at the ORNL were never as organized or as extensive as studies done in terrestrial ecosystems. The aquatic work began with a study by Louis Krumholz, a fisheries biologist from the Tennessee Valley Authority. Krumholz was commissioned by the Health Physics Department to carry out an ecological survey of White Oak Lake from Fall 1950 to Spring 1953 before it was partially drained. This study focused on fishes and it included radioassay with a Geiger-Mueller counter. The lake was found to have a normal composition of fish species with a dominance of bluegills (*Lepomis* M. *macrochrus*) and black crappie (*Pomixis nigro-maculatus*) along with eight other species. A large standing crop of fish biomass was found, probably, as was noted, because the lake was eutrophic from inputs of nutrients from the ORNL sewage treatment plant. Krumholz (1956) suggested that the lake "was overpopulated with bluegills" which had "resulted in a stunting of the entire bluegill population". A full food chain study was not carried out, but some samples of algae, zooplankton and insect larvae were included in the radioassay. Most radioactivity in the fishes was found in the bones. Internal dose was not estimated but it was expected to be several times the external dose. No mutations were found in any of the fishes, but the effects of irradiation were suspected to have shortened the fishes' lifespan and slowed their growth rate compared to other aquatic systems in the area.

Most of the aquatic ecology research at ORNL related, directly or indirectly, to the risk posed by the releases of radionuclides into the Clinch River from White Oak Creek. Daniel J. Nelson was hired in 1959 to lead the aquatic research. An extensive research program on radioactivity in the river was carried out by a consortium including the ORNL, the Tennessee Valley Authority, the U.S. Public Health Service and the U.S. Geological Survey. This work lasted seven years and it resulted in a number of studies. Dilution of the radionuclides, for example, was found to be effective, and 95% of the radioactivity in the Clinch River sediments was found within 24 kilometers (15 miles) downstream from the confluence with White Oak Creek (Parker et al. 1966, Pickering et al. 1966). The overall result of the study was that the risk to humans who utilized river water was below permissible limits (Cowser et al. 1966). Nelson's most important contribution to the Clinch River study was his research on clams as biological indicators of radionuclides. Fifteen species of clams (family Unionidae) were studied in the Clinch and Tennessee Rivers by measuring the concentration of Sr-90 in their shells. These shells are essentially pure $CaCO_3$, and Sr-90 was found to substitute for calcium in low concentrations. The shell is deposited in distinct annual rings, so the chemical composition of the shell represents a history of deposition of strontium. Because the clams are relatively immobile, they are good indicators for water chemistry in the locations where they live. Evidence of radiostrontium was found in clam shells up to nearly 800 kilometers (500 miles) downstream from the outfall of White Oak Creek into the Clinch River, but the total mass of the radionuclides in river clams was small due to dilution (Nelson 1962, 1963, 1964).

Nelson and his group also carried out research on fishes (Nelson 1969, Reed and Nelson 1969), crayfish (Wiser and Nelson 1964, Schurr and Stamper 1962) and snails (Malone and Nelson 1969, Cooley 1973) as part of the Clinch River study and other aquatic ecology assessments.

As part of the aquatic assessments of radioactivity in White Oak Creek, B. G. Blaylock studied cytogenetics in midge fly larvae (*Chironomus tetans*) over a five-year period (1960–1964) as part of his Ph.D. dissertation work at the University of Tennessee. He estimated radiation exposures to the midge larvae, which live in tubes in the bottom muds, to be approximately 230 R per year (Nelson and Blaylock 1963). This was 1,000 times greater than background radiation of control populations living in a nearby uncontaminated stream. A number of chromosomal aberrations were found in the larvae living in the contaminated muds relative to the control larvae and he concluded, "There is little doubt that this population has suffered genetic damage from the high radiation background" (Blaylock 1966a). However, he also suggested that the population, which had passed through 130 generations (22 years) since radioactive wastes had been discharged into the creek, did not exhibit reduced evolutionary fitness because the new aberrations were eliminated by natural selection or genetic drift (Blaylock 1965, 1966a, 1966b). Blaylock (1969) later studied embryos and brood size of mosquito fish (*Gambusia affinis affinis*) from White Oak Lake and a nearby uncontaminated pond as part of the ecological assessment. The mosquito fish from White Oak Lake had a higher frequency of dead embryos and embryos with abnormalities, but their brood size was actually higher in comparison with the control population. Blaylock (1969) concluded:

> These results suggest that an increased fecundity is a means by which a natural population of organisms having a relatively short life cycle and producing a large number of progeny can adjust rapidly to an increased mortality caused by radiation.

Overall, the research by Blaylock on White Oak Lake represents a significant contribution from the aquatic ecology group at ORNL that demonstrates how processes can adjust population structure and dynamics in respect to radiation exposure. The organisms he studied, midge larvae and mosquito fish, are well known to be hardy, adaptable species and their ability to exist in a contaminated environment is another example of the self-organizing properties of ecosystems. Organisms such as these survive and adjust their populations, thereby contributing to the larger ecosystem.

Bernard Patten participated in aquatic ecology at the ORNL in addition to his work in ecosystem modeling. Patten was hired in 1964 with "dual capacity" in aquatic ecology and the newly formed theoretical/systems ecology group. He came to the ORNL with a record of highly theoretical analyses, mostly based on fieldwork he had done on estuarine phytoplankton communities (Patten 1959, 1961). At Oak Ridge he carried out fieldwork on algal growth and radioactivity (Patten and Iverson 1966, Neal et al. 1967) and he continued to develop theoretical analyses using his past research (Patten and Van Dyne 1968, Patten and Chabot 1966). By the mid-1960s the Oak Ridge Radioecology Group had grown in size, essentially to a critical mass, with much new thinking on how to analyze data that was being gathered from ecosystems contaminated with radioactivity at the ORNL and from undisturbed ecosystems in general. One of the new approaches was Systems Ecology. Patten helped catalyze the field with Jerry Olson and another recent hire, George Van Dyne, as one by-product of the Oak Ridge Radioecology Group's collaboration (Patten 1966).

Heritage

By the end of the 1960s the radioecology group had grown to more than 80 scientists and support staff (Bocking 1997). This growth was in large part due to Auerbach's skill as an administrator and to the group's research output. Auerbach had become a leader in the field of Radioecology (Auerbach 1965, 1973, 1984), but interest in this field declined as new sources of research funding became available and as social awareness of other environmental problems had grown, as reflected by the National Environmental Policy Act of 1969.

While issues with radioactive wastes continued at the ORNL, the ecologists broadened their focus and shifted to other topics, both applied and theoretical. Auerbach became president of the Ecological Society of America in 1971 and his roles as an administrator expanded beyond Oak Ridge (Reichle 1986). The radioecology group was renamed the Environmental Sciences Division in 1970 as a reflection of the changes that were occurring.

The coming of the International Biological Program (IBP) was an opportunity for the ORNL ecologists to translate the knowledge about ecosystems that they had developed from radioecology into a broader framework. The IBP was a massive international program of focused, collaborative research about many biological topics, including ecosystem structure and function. As noted by Botkin (1977), "The IBP was conceived as an ecological analog of the International Geophysical Year, with the biological basis of productivity and human welfare as a theme." It lasted from 1966 to 1976 and included large-scale funding by the U.S. National Science Foundation for research in six biomes. The Eastern Deciduous Forest was one of the six U.S. biomes. Auerbach became the director, administering its research, which spanned five sites: Duke Forest and the Triangle Research site in North Carolina, the Lake George basin in New York, the Lake Wingra basin in Wisconsin and the Coweeta Hydrologic Laboratory in Georgia (Auerbach 1971). At Oak Ridge a natural transition occurred from Radioecology to general ecology of forested ecosystems and the AEC contributed funding to the transition. Work in the Cesium Forest was concluded and a similar forest on the ORNL grounds became one of the major sites in the IBP's Eastern Deciduous Forest Biome with research shifting from tracking radionuclides to tracking carbon flows (Reichle et al. 1973, Harris et al. 1975, Sollins et al. 1976). In essence, the thinking and methods developed in the old Cesium Forest, with research funded by the AEC (see Figure 6.3), took on a new life as measurements of carbon flows were made, forming the basis for a detailed budget (Figure 6.6). David Reichle helped synthesize much of the Eastern Deciduous Forest Biome's work (Reichle 1975a, 1975b, 1981). Of note, Reichle's original work on the cryptozoa and Martin Witkamp's work on microbes in the Cesium Forest became the basis for an expanded model of the functioning of the forest floor within the larger ecosystem (Reichle 1971, Witkamp 1971b, Edwards et al. 1970, Edwards and Harris 1977). New researchers, such as Hank Shugart, arrived near the end of the IBP work and continued to develop the forest ecosystem

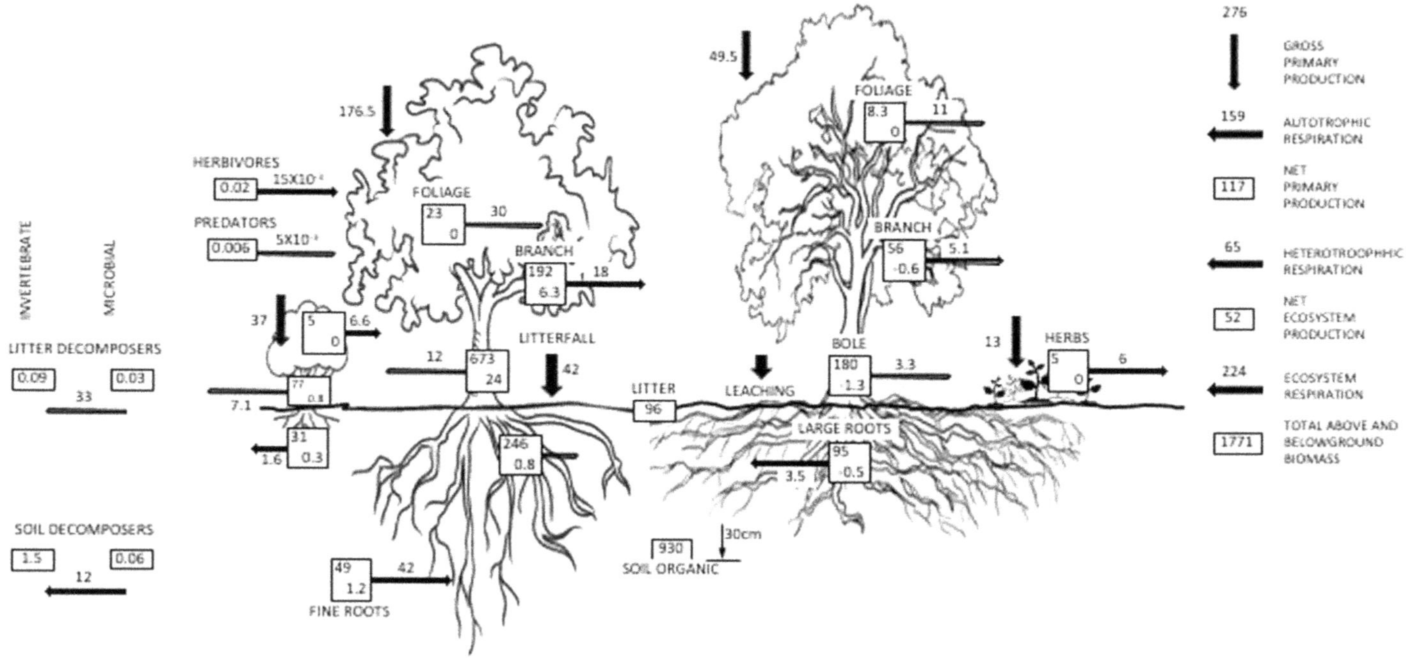

FIGURE 6.6 Model of the carbon cycle in the mesic hardwood forest at the Oak Ridge National Laboratory. Units are 10^5 grams per hectare for storages and 10^5 grams per hectare per year for flows.

Source: From Reichle et al. 1973.

model (Shugart 1984, Shugart et al. 1974) that had been started in the Cesium Forest. In general, the IBP played an important role in the development of the ecosystem concept (Coleman 2010) and the heritage of the Oak Ridge ecologists contributed significantly to this development. Their conception of the ecosystem changed from their original views of White Oak Lake Bed to those of the IBP work (Reichle 1975a, Reichle and Auerbach 1972, Riechle et al. 1975). Of historical note, the IBP was a kind of precursor for the Long Term Ecological Research (LTER) Program that began in the 1990s (Hobbie et al. 2003), but Oak Ridge was not chosen as one of the LTER sites—though if Stanley Auerbach had still been active, it probably would have been!

When Radioecology studies declined as a central focus at Oak Ridge in the late 1960s, another shift in research took place, to work on the global carbon cycle, which was getting attention as the causal basis for the greenhouse effect and, ultimately, climate change. Jerry Olson began scaling up data on forest carbon cycles and he contributed some foundations for future research (Olson 1970, 1975). In this regard, Dale et al. (2021) suggest Olson was "far ahead of most in his understanding of the role of the biosphere in the global carbon cycle". Oak Ridge scientists have continued global carbon cycle work (see, for example, Trabalka et al. 1986, Dale 1994, Post et al. 1990), which is ongoing.

Much of the heritage of the Oak Ridge Radioecology group involves individuals, some who stayed and some who left for other institutions. Many individuals who worked at Oak Ridge with support from the AEC went on to be leaders in the field of Ecology throughout their careers.

In addition to those already covered above, many others continued ecological research after the importance of Radioecology declined, but only a few of them can be covered here. Daniel Nelson continued aquatic ecology research by initiating a watershed project on the 97.5 hectares (241 acres) Walker Branch stream drainage network in the ORNL reservation. This project was more or less modeled on the Hubbard Brook Watershed study with a focus on studying hydrologic and chemical inputs and outputs (Nelson 1970, Elwood and Henderson 1975). Nelson died in 1975, but the research on Walker Branch has grown into an important site for watershed research (Johnson and Van Hook 1989).

Robert O'Neill spent his entire career at Oak Ridge and became a leader in ecological modeling and theoretical ecology (see, for example, O'Neill and Reichle 1980, O'Neill et al., 1986). His accomplishments seem striking given his somewhat narrow original focus on millipede ecology! He ultimately came to express doubts about ecosystem science, at least rhetorically (O'Neill 2001). Near the end of his career he became interested in the transdisciplinary field of ecological economics (O'Neill 1996, O'Neill and Kahn 2000).

Several of the scientists of the ORNL Radioecology group left to continue their careers and to make important contributions to Ecology elsewhere. George Van Dyne spent only three years at Oak Ridge, but while there he made significant contributions to the system ecology program (see, for example, Van Dyne 1966). He left in 1967 for Colorado State University, where he became the leader of the

Grassland Biome project of the IBP (Coleman et al. 2004a). Auerbach (1995) wrote a summary of his career, which was cut short by his death in 1981.

D. A. Crossley Jr., Auerbach's first hire into the Radioecology group, left Oak Ridge for the University of Georgia. He continued work on invertebrate ecology there, producing a number of review articles with his graduate students (Crossley 1977, Gist and Crossley 1975, Seastedt and Crossley 1984) along with coauthoring a text on soil ecology (Coleman et al. 2004b).

The research carried out by B. C. Patten during his five years at the ORNL made a significant impact on his future work. Patten left Oak Ridge in 1969 for the University of Georgia, where he continued the work on systems ecology that he had initially developed through his collaboration with Jerry Olson and George Van Dyne. Of note, Patten edited a series of books on "Systems Analysis and Simulation in Ecology" that synthesized and stimulated the field of Systems Ecology. Over the next 50 years he developed his own expression of the field based on network analysis (Patten 1998, 2016, Patten et al. 1976). The lasting impact of his microcosm studies, carried out at Oak Ridge, as described in the following section, is an example of what is meant in this book by the heritage of Radioecology for the history of ecology:

> My work with Martin Witkamp was a turning point in my recognition of the power of networks to influence organisms. In laboratory microcosms, Witkamp demonstrated, using a radioactive tracer, that changes only in the ways organisms were interconnected changed their function, right down to physiological rates of feeding and excretion—organism properties, not community properties. I lectured during that period on what I called "the network variable in ecology", trying to say what should have been an obvious corollary of everyone's definition of system as a partially interconnected set of component parts . . .: change the network, change the system—and the things and their processes within it. Nothing I have learned since has deflected me from this conclusion; all observation and study have reinforced it.
>
> *(Patten 1991)*

In addition to his own work, the many students that Patten trained in network analysis over his career (see the "academic tree" in Borrett et al. 2014) are a record of his intellectual network, which is partly due to his formative time of the 1960s in the Oak Ridge Radioecology Group.

Literature Cited

Auerbach, S. I. 1949. A preliminary ecological study on certain deciduous forest centipedes. American Midland Naturalist 42:220–227.

Auerbach, S. I. 1951. The centipedes of the Chicago area with special reference to their ecology. Ecological Monographs 21:97–124.

Auerbach, S. I. 1958. The soil ecosystem and radioactive waste disposal to the ground. Ecology 39:522–529.

Auerbach, S. I. 1959. Ecological Research: White Oak Lake Bed Studies. Health Physics Division Annual Progress Report for Period Ending July 31, 1959. ORNL-2806, Oak Ridge National Laboratory, Oak Ridge, TN.

Auerbach, S. I. 1965. Radionuclide cycling: Current status and future needs. Health Physics 11:1355–1361.

Auerbach, S. I. 1971. The deciduous forest biome programme in the United States of America. Pp. 677–684. In: P. Duvigneaud (ed.). Productivity of Forest Ecosystems. UNESCO, Paris.

Auerbach, S. I. 1973. Contribution of radioecology to AEC mission programs. Pp. 3–8. In: D. J. Nelson (ed.). Radionuclides in Ecosystems. US AEC, Washington, DC.

Auerbach, S. I. 1984. A perspective on radioecological research. Journal of the Society of Radiological Protection 4:100–105.

Auerbach, S. I. 1995. George M. Van Dyne—A reminiscence. Pp. xxvii–xxx. In: B. C. Patten and S. E. Jorgensen (eds.). Complex Ecology: The Part-Whole Relation in Ecosystems. Prentice Hall, Englewood Cliffs, NJ.

Auerbach, S. I., D. A. Crossley, Jr. and M. D. Engelman. 1957. Effects of gamma radiation on *Collembola* population growth. Science 126:614.

Auerbach, S. I. and J. S. Olson. 1963. Biological and environmental behavior of Ruthenium and Rhodium. Pp. 509–520. In: V. Schultz and A. W. Klement, Jr. (eds.). Radioecology. Reinhold Publ. Corp., New York, NY.

Auerbach, S. I., J. S. Olson and H. D. Waller. 1964. Landscape studies using 137Cs. Nature 201:761–764.

Auerbach, S. I., J. S. Olson and H. D. Waller. 1967. Data on activity transfer to the forest floor. Pp. 467–469. In: B. Åberg and F. P. Hungate (eds.). Radioecological Concentration Processes. Pergamon Press, Oxford, UK.

Blaylock, B. G. 1965. Chromosomal aberrations in a natural population of *Chironomus tentans* exposed to chronic low-level radiation. Evolution 19:421–429.

Blaylock, B. G. 1966a. Cytogenetic study of natural population of *Chironomus* inhabiting an area contaminated by radioactive waste. Pp. 835–846. In: Disposal of Radioactive Wastes into Seas, Oceans and Surface Waters. International Atomic Energy Agency, Vienna.

Blaylock, B. G. 1966b. Chromosomal polymorphism in irradiated natural populations of *Chironomus*. Genetics 53:131–136.

Blaylock, B. G. 1969. The fecundity of a *Gambusia affinis affinis* population exposed to chronic environmental radiation. Radiation Research 37:108–117.

Bledsoe, L. J. and J. S. Olson. 1970. COMSYS1: A stepwise compartmental simulation program. ORNL—TM—2413, Oak Ridge National Laboratory, Oak Ridge, TN.

Bocking, S. 1995. Ecosystems, ecologists, and the atom: Environmental research at Oak Ridge National Laboratory. Journal of the History of Biology 28:1–47.

Bocking, S. 1997. Ecologists and Environmental Politics: A History of Contemporary Ecology. Yale University Press, New Haven, CT. (Chapter 4).

Bocock, K. L. and O. J. Gilbert. 1957. The disappearance of leaf litter under different woodland conditions. Plant and Soil 9:179–185.

Borrett, S. R., B. D. Fath and S. J. Whipple. 2014. Introduction to the special issues "Systems Ecology: A Network Perspective and Retrospective". Ecological Modelling 293:1–3.

Botkin, D. B. 1977. Bits, bytes, and IBP. Bioscience 27:385.

Childs, J. E., Jr. and G. E. Cosgrove. 1966. A study of pathological conditions in wild rodents in radioactive areas. American Midland Naturalist 76:309–324.

Claus, W. D. 1958. What is health physics? Health Physics 1:56–61.
Coleman, D. C. 2010. Big Ecology: The Emergence of Ecosystem Science. University of California Press, Berkeley, CA.
Coleman, D. C., D. A. Crossley, Jr. and P. F. Hendrix. 2004b. Fundamentals of Soil Ecology. Elsevier, Amsterdam.
Coleman, D. C., D. M. Swift and J. E. Mitchell. 2004a. From the frontier to the biosphere: A brief history of the USIBP Grassland Biome program and its impacts on scientific research in North America. Rangelands 26:8–15.
Cooley, J. L. 1973. Effects of chronic environmental radiation on a natural population of the aquatic snail Physa heterostropa. Radiation Research 54:130–140.
Cowser, K. E. and F. L. Parker. 1958. Soil disposal of radioactive wastes at ORNL: Criteria and techniques of site selection and monitoring. Health Physics 1:152–163.
Cowser, K. E., W. S. Snyder, C. P. McCammon, C. P. Straub, O. W. Kochtitzky, R. L. Hervin, E. G. Struxness and R. J. Morton. 1966. Evaluation of radiation dose to man from radionuclides released to the Clinch River. Pp. 639–668. In: Disposal of Radioactive Wastes into Seas, Oceans and Surface Waters. International Atomic Energy Agency, Vienna.
Croker, R. A. 1991. Pioneer Ecologist: The Life and Work of Victor Ernest Shelford, 1877–1968. Smithsonian Institution Press, Washington, DC.
Crossley, D. A., Jr. 1963a. Use of radioactive tracers in the study of insect-plant relationships. Pp. 43–54. In: Radiation and Radioisotopes Applied to Insects of Agricultural Importance. International Atomic Energy Agency, Vienna.
Crossley, D. A., Jr. 1963b. Consumption of vegetation by insects. Pp. 427–430. In: V. Schultz and A. W. Klement, Jr. (eds.). Radioecology. Reinhold Publ. Co., New York, NY.
Crossley, D. A., Jr. 1966. Radioisotope measurement of food consumption by a leaf beetle species, *Chrysomela knabi* Brown. Ecology 47:1–8.
Crossley, D. A., Jr. 1969. Comparative movement of 106Ru, 60Co, and 137 Cs in arthropod food chains. Pp. 687–697. In: D. J. Nelson and F. C. Evans (eds.). Symposium on Radioecology. CONF-670503. US AEC, Washington, DC.
Crossley, D. A., Jr. 1977. The roles of terrestrial saprophagous arthropods in forest soils: Current status of concepts. Pp. 49–56. In: W. J. Mattson (ed.). The Role of Arthropods in Forest Ecosystems. Springer-Verlag, New York, NY.
Crossley, D. A., Jr. and M. P. Hoglund. 1962. A litter-bag method for the study of microarthropods inhabiting leaf litter. Ecology 43:571–573.
Crossley, D. A., Jr. and H. F. Howden. 1961. Insect-vegetation relationships in an area contaminated by radioactive wastes. Ecology 42: 302–317.
Crossley, D. A., Jr. and M. E. Pryor. 1960. The uptake and elimination of cesium-137 by a grasshopper—*Romalea microptera*. Health Physics 4:16–20.
Crossley, D. A., Jr. and J. H. Schnell 1961. Comparative elimination of radiocesium and radiostrontium by grasshoppers. Annals of the Entomological Society of America 54:459–461.
Dahlman, R. C., S. I. Auerbach and P. B. Dunaway. 1969. Behaviour of 137Cs-tagged particles in a fescue meadow. Pp. 153–165. In: Environmental Contamination by Radioactive Materials. International Atomic Energy Agency, Vienna.
Dale, V. H. 1994. Terrestrial CO2 flux: The challenge of interdisciplinary research. Pp. 1–14. In: V. H. Dale (ed.). Effects of Land-Use Change on Atmospheric CO2 Concentrations, South and Southeast Asia as a Case Study. Springer-Verlag, New York, NY.
Dale, V. H., M. Post and R. J. Norby. 2021. Resolution of respect, Jerry S. Olson (1928–2021). Bulletin of the Ecological Society of America 102:1–8.
Davis, J. J. and R. F. Foster. 1958. Bioaccumulation of radioisotopes through aquatic food chains. Ecology 39:530–535.

DeSelm, H. R. and R. E. Shanks. 1961. The ecology of White Oak Lake Bed: A radioactive waste area Part II. Pp. 1359–1363. In: Recent Advances in Botany, Volume 2. University of Toronto Press, Toronto, Canada.

DeSelm, H. R. and R. E. Shanks. 1963. Accumulation and cycling of organic matter and chemical constituents during early vegetational succession on a radioactive waste disposal area. Pp. 83–96. In: V. Schultz and A. W. Klement, Jr. (eds.). Radioecology. Reinhold Publ. Co., New York, NY.

DeSelm, H. R. and R. E. Shanks. 1967. Vegetation and floristic changes on a portion of White Oak Lake bed. Ecology 48:419–425.

Dunaway, P. B. and G. E. Cosgrove. 1966. A skin carcinoma in a cotton rat. American Midland Naturalist 75:535–537.

Dunaway, P. B. and S. V. Kaye. 1961. Studies of small-mammal populations on the radioactive White Oak Lake bed. Transactions of the North American Wildlife Conference 29:166–186.

Dunaway, P. B. and S. V. Kaye. 1963. Effects of ionizing radiation on mammal populations on the White Oak Lake bed. Pp. 333–338. In: V. Schultz and A. W. Klement, Jr. (eds.). Radioecology. Reinhold Publ. Co., New York, NY.

Dunaway, P. B. and S. V. Kaye. 1964. Weights of cotton rates in relation to season, breeding, and environmental radioactive contamination. American Midland Naturalist 71:141–155.

Dunaway, P. B., L. L. Lewis, J. D. Story, J. A. Payne and J. M. Inglis. 1969. Radiation effects in the Soricidae, Cricetidae, and Muridae. Pp. 173–184. In: D. J. Nelson and F. C. Evans (eds.). Symposium on Radioecology. CONF-670503, US AEC, Washington, DC.

Edwards, C. A., D. E. Reichle and D. A. Crossley, Jr. 1970. The role of soil invertebrates in turnover of organic matter and nutrients. Pp. 147–172. In: D. E. Reichle (ed.). Analysis of Temperate Forest Ecosystems. Springer-Verlag, New York, NY.

Edwards, N. T. and W. F. Harris. 1977. Carbon cycling in a mixed deciduous forest floor. Ecology 58:431–437.

Elwood, J. W. and G. S. Henderson. 1975. Hydrologic and chemical budgets at Oak Ridge, Tennessee. Pp. 31–50. In: A. D. Hasler (ed.). Coupling of Land and Water Systems. Springer-Verlag, New York, NY.

Engelmann, M. 1970. Resolution of respect: Orlando Park, 1901–1969. Bulletin of the Ecological Society of America 51:16–20.

Evans, F. C. 1956. Ecosystem as the basic unit in ecology. Science 123:1127–1128.

Fenton, G. R. 1947. The soil fauna: With special reference to the ecosystem of the forest soil. Journal of Animal Ecology 16:76–93.

Gist, C. S. and D. A. Crossley, Jr. 1975. A model of mineral-element cycling for an invertebrate food web in a southeastern hardwood forest litter community. Pp. 84–106. In: F. G. Howell, J. B. Gentry and M. H. Smith (eds.). Mineral Cycling in Southeastern Ecosystems. US Energy Research and Development Administration CONF-740513, Washington, DC.

Graham, B. F., Jr. 1954. A technique for introducing radioactive isotopes into tree stems. Ecology 35:415.

Graham, E. R. 1958. Uptake of waste Sr 90 and Cs 137 by soil and vegetation. Soil Science 86:91–97.

Harris, W. F., P. Sollins, N. T. Edwards, B. E. Dinger and H. H. Shugart. 1975. Analysis of carbon flow and productivity in a temperate deciduous forest ecosystem. pp. 116–122. In: Productivity of World Ecosystems. National Academy of Sciences, Washington, DC.

Hobbie, J., S. R. Carpenter, N. B. Grimm, J. R. Gosz and T. R. Seastedt. 2003. The US long-term ecological research program. Bioscience 53:21–32.

Hoffman, G. R. 1972. The accumulation of Cesium-137 by cryptogam in a *Liriodendron tulipifera* forest. Botanical Gazette 133:107–119.

Howden, H. F. 1957. Investigation on sterility and deformities of *Onthophagus* (Coleoptera, Scarabaeidae) induced by gamma radiation. Annals of the Entomological Society of America 50:1–9.

Howden, H. F. and S. I. Auerbach. 1958. Some effects of gamma radiation on *Trogoderma sternale* Jayne. Annals of the Entomological Society of America 51:48–51.

Jenny, H. 1941. Factors of Soil Formation: A System of Quantitative Pedology. McGraw-Hill Book Co., New York, NY.

Johnson, C. W. and C. O. Jackson. 1981. City Behind a Fence, Oak Ridge, Tennessee 1942–1946. University of Tennessee Press, Knoxville, TN.

Johnson, D. W. and R. I. Van Hook (eds.). 1989. Analysis of Biogeochemical Cycling Processes in Walker Branch Watershed. Springer-Verlag, New York, NY.

Kaye, S. V. and P. B. Dunaway. 1962. Bioaccumulation of radioactive isotopes by herbivorous small mammals. Health Physics 7:205–217.

Kaye, S. V. and P. B. Dunaway. 1963. Estimation of dose rate and equilibrium state from bioaccumulation of radionuclides by mammals. Pp. 107–111. In: V. Schultz and A. W. Klement, Jr. (eds.). Radioecology. Reinhold Publ. Co., New York, NY.

Kitchings, J. T. III, P. B. Dunaway, J. D. Story and L. E. Tucker. 1968. Use of radioiron (59Fe) as an index to hemopoietic damage caused by ionizing radiation. Journal of the Tennessee Academy of Science 43:85–87.

Krause, C. 1992. Oak Ridge national laboratory: The first 50 years. The ORNL Review 25(3 and 4).

Krumholz, L. A. 1956. Observations on the fish population of a lake contaminated by radioactive wastes. Bulletin Volume 110: Article 4. American Museum of Natural History, New York, NY.

Kwa, C. 1993. Radiation ecology, systems ecology and the management of the environment. Pp. 213–249. In: M. Shortland (ed.). Science and Nature, Essays in the History of the Environmental Sciences. British Society for the History of Science. Oxford, UK.

Lomenick, T. F. and D. A. Gardiner. 1965. The occurrence and retention of radionuclides in the sediments of White Oak Lake. Health Physics 11:567–577.

Malone, C. R. and D. J. Nelson. 1969. Feeding rates of freshwater snails (*Goniiobasis clavaeformis*) determined with cobalt 60. Ecology 50:728–730.

McBrayer, J. F. and D. E. Reichle. 1971. Trophic structure and feeding rates of forest soil invertebrate populations. Oikos 22:381–388.

Morgan, K. Z. and K. M. Peterson. 1999. The Angry Genie, One Man's Walk Through the Nuclear Age. University of Oklahoma Press, Norman, OK.

Morton, R. J. and E. G. Struxness. 1956. Ground disposal of radioactive wastes. American Journal of Public Health 46:156–163.

Moulder, B. C. and D. E. Reichle. 1972. Significance of spider predation in the energy dynamics of forest-floor arthropod communities. Ecological Monographs 42:473–498.

Neal, E. C., B. C. Patten and C. E. DePoe. 1967. Periphyton growth on artificial substrates in a radioactively contaminated lake. Ecology 48:918–924.

Neel, R. B. and J. S. Olson. 1962. Use of analog computers for simulating the movement of isotopes in ecosystems. US AEC Report ORNL-3172, Oak Ridge National Laboratory.

Nelson, D. J. 1962. Clams as indicators of strontium-90. Science 137:38–39.

Nelson, D. J. 1963. The strontium and calcium relationship in Clinch and Tennessee River mollusks. Pp. 203–211. In: V. Schultz and A. W. Klement, Jr. (eds.). Radioecology. Reinhold Publ. Co., New York, NY.

Nelson, D. J. 1964. Biological vectors and reservoirs of Strontium-90. Nature 203:420.

Nelson, D. J. 1969. Cesium, Cesium-137, and potassium concentrations in white crappie and other Clinch River fish. Pp. 240–248. In: D. J. Nelson and F. C. Evans (eds.). Symposium on Radioecology, CONF-670503. US AEC, Washington, DC.

Nelson, D. J. 1970. Measurement and sampling of outputs from watersheds. Pp. 257–267. In: D. E. Reichle (ed.). Analysis of Temperate Forest Ecosystems. Springer-Verlag, New York, NY.

Nelson, D. J. and B. G. Blaylock. 1963. The preliminary investigation of salivary gland chromosomes of *Chironomus tetans* Fabr. From the Clinch River. Pp. 367–372. In: V. Schultz and A. W. Klement, Jr. (eds.). Radioecology. Reinhold Publ. Co., New York, NY.

Olson, J. S. 1958. Rates of succession and soil changes on southern Lake Michigan sand dunes. Botanical Gazette 119:125–170.

Olson, J. S. 1963a. Analog computer models for movement of nuclides through ecosystems. Pp. 121–125. In: V. Schultz and A. W. Klement, Jr. (eds.). Radioecology. Reinhold Publ. Co., New York, NY.

Olson, J. S. 1963b. Energy storage and the balance of producers and decomposers in ecological systems. Ecology 44:322–331.

Olson, J. S. 1965. Equations for cesium transfer in a *Liriodendron* forest. Health Physics 11:1385–1392.

Olson, J. S. 1970. Carbon cycles and temperate woodlands. Pp. 226–241. In: D. E. Reichle (ed.). Analysis of Temperate Forest Ecosystems. Springer-Verlag. New York, NY.

Olson, J. S. 1975. Productivity of forest ecosystems. Pp. 33–43. In: Productivity of World Ecosystems. National Academy of Sciences, Washington, DC.

Olson, J. S. and D. A. Crossley, Jr. 1963. Tracer studies of the breakdown of forest litter. Pp. 411–416. In: V. Schultz and A. W. Klement, Jr. (eds.). Radioecology. Reinhold Publ. Co., New York, NY.

O'Neill, R. V. 1967. Niche segregation in seven species of diplopods. Ecology 48:983.

O'Neill, R. V. 1968. Population energetics of the millipede, *Narceus americanus* (Beauvois). Ecology 49:803–809.

O'Neill, R. V. 1971. Systems approaches to the study of forest floor arthropods. Pp. 441–477. In: B. C. Patten (ed.). Systems Analysis and Simulation in Ecology. Academic Press, New York, NY.

O'Neill, R. V. 1996. Perspectives on economics and ecology. Ecological Applications 6:1031–1033.

O'Neill, R. V. 2001. Is it time to bury the ecosystem concept? (with full military honors, of course!). Ecology 82:3275–3284.

O'Neill, R. V., D. L. DeAngelis, J. B. Waide and T. F. H. Allen. 1986. A Hierarchical Concept of Ecosystems. Princeton University Press, Princeton, NJ.

O'Neill, R. V. and J. R. Kahn. 2000. *Homo economus* as a keystone species. Bioscience 50:333–337.

O'Neill, R. V. and D. E. Reichle. 1980. Dimensions of ecosystem theory. Pp. 11–26. In: R. H. Waring (ed.). Forests: Fresh Perspective from Ecosystem Analysis. Oregon State University Press, Corvallis, OR.

Park, O. and S. Auerbach. 1954. Further study of the tree-hole complex with emphasis on quantitative aspects of the fauna. Ecology 35:208–222.

Parker, F. L., M. A. Churchill, R. W. Andrew, B. J. Frederick, P. H. Carrigan, Jr., J. S. Cragwall, R., S. L. Jones, E. G. Struxnew and R. J. Morton. 1966. Dilution, dispersion and mass transport of radionuclides in the Clinch and Tennessee Rivers. Pp. 33–55. In: Disposal of Radioactive Wastes into Seas, Oceans and Surface Waters. International Atomic Energy Agency, Vienna.

Patten, B. C. 1959. An introduction to the cybernetics of the ecosystem: The trophic-dynamic aspect. Ecology 40:221–231.

Patten, B. C. 1961. Negentropy flow in communities of plankton. Limnology and Oceanography 6:26–30.

Patten, B. C. 1966. Systems ecology: A course sequence in mathematical ecology. Bioscience 16:593–598.
Patten, B. C. 1991. Network ecology: Indirect determination of the life-environment relationship in ecosystems. Pp. 288–351. In: M. Higashi and T. Burns (eds.). Theoretical Studies of Ecosystems: The Network Perspective. Cambridge University Press, Cambridge, UK.
Patten, B. C. 1998. Network orientors: Steps toward a cosmography of ecosystems: Orientors for directional development, self-organization, and autoevolution. Pp. 137–160. In: F. Muller and M. Leupelt (eds.). Eco Targets, Goal Functions, and Orientors. Springer-Verlag, Berlin.
Patten, B. C. 2016. The cardinal hypotheses of Holoecology: Facets for a general systems theory of the organism-environment relationship. Ecological Modelling 319:63–111.
Patten, B. C., R. W. Bosserman, J. T. Finn and W. G. Cale. 1976. Propagation of cause in ecosystems. P. 457. In: B. C. Patten (ed.). Systems Analysis and Simulation in Ecology, Volume 4. Academic Press, New York, NY.
Patten, B. C. and B. F. Chabot. 1966. Factorial productivity experiments in a shallow estuary: Characteristics of response surfaces. Chesapeake Science 7:117–136.
Patten, B. C. and R. L. Iverson. 1966. Photosynthesis and uptake of strontium=85 in freshwater plankton. Nature 211:96–97.
Patten, B. C. and G. M. Van Dyne. 1968. Factorial productivity experiments in a shallow estuary: Energetics of individual plankton species in a mixed population. Limnology and Oceanography 13:309–314.
Patten, B. D. and M. Witkamp. 1967. Systems analysis of 134 Cesium kinetics in terrestrial microcosms. Ecology 48:813–824.
Pickering, R. J., P. H. Carrigan, Jr., T. Tamura, H. H. Abee, J. W. Beverage and R. W. Andrew, Jr. 1966. Radioactivity in bottom sediments of the Clinch and Tennessee Rivers. Pp. 57–88. In: Disposal of Radioactive Wastes into Seas, Oceans and Surface Waters. International Atomic Energy Agency, Vienna.
Plummer, G. L., D. A. Crossley, Jr. and D. A. Gardiner. 1965. Growth response of two sedges inhabiting a radioactive waste disposal area. Ecology 46:346–348.
Post, W. M., T.-H. Peng, W. R. Emanuel, A. W. King, V. H. Dale and D. L. DeAngelis. 1990. The global carbon cycle. American Scientist 78:310–326.
Reed, J. R. and D. J. Nelson. 1969. Radiostrontium uptake in blood and flesh in bluegills (*Lepomis macrochirus*). Pp. 226–233. In: D. J. Nelson and F. C. Evans (eds.). Symposium on Radioecology, CONF-670503. US AEC, Washington, DC.
Reichle, D. E. 1964. The population ecology, distribution and abundance of bog-inhabiting *Pselaphid* beetles. PhD Dissertation, Northwestern University, Evanston, IL.
Reichle, D. E. 1967. Radioisotope turnover and energy flow in terrestrial isopod populations. Ecology 48:351–366.
Reichle, D. E. 1968. Relation of body size to food intake, oxygen consumption, and trace element metabolism in forest floor arthropods. Ecology 49:538–542.
Reichle, D. E. 1969. Measurement of elemental assimilation by animals from radioisotope retention patterns. Ecology 50:1102–1104.
Reichle, D. E. 1971. Energy and nutrient metabolism of soil and litter invertebrates. Pp. 465–477. In: P. Duvigneaud (ed.). Productivity of Forest Ecosystems. UNESCO, Paris.
Reichle, D. E. (ed.). 1975a. Analysis of Temperate Forest Ecosystems. Springer-Verlag, New York, NY.
Reichle, D. E. 1975b. Advances in ecosystem analysis. Bioscience 25:257–264.
Reichle, D. E. (ed.). 1981. Dynamic Properties of Forest Ecosystems. Cambridge University Press, Cambridge, UK.

Reichle, D. E. 1986. Distinguished service citation, Stanley I. Auerbach. Bulleting of the Ecological Society of America 67:44–45.
Reichle, D. E. and S. I. Auerbach. 1972. Analysis of ecosystems. Pp. 260–280. In: J. A. Behnke (ed.). Challenging Biological Problems, Directions Toward their Solution. Oxford University Press, New York, NY.
Reichle, D. E. and S. I. Auerbach. 2003. U.S. Radioecology Research Programs of the Atomic Energy Commission in the 1950s. ORNL/TM-2003/280, Oak Ridge National Laboratories, Oak Ridge, TN.
Reichle, D. E. and D. A. Crossley, Jr. 1965. Radiocesium dispersion in a cryptozoan food web. Health Physics 11:1375–1384.
Reichle, D. E. and D. A. Crossley, Jr. 1967. Investigation on heterotrophic productivity in forest insect communities. Pp. 563–587. In: K. Petrusewicz (ed.). Secondary Productivity of Terrestrial Ecosystems. Polish Academy of Science, Warszawa, Krakow.
Reichle, D. E. and D. A. Crossley, Jr. 1969. Trophic level concentrations of cesium-137, sodium, and potassium in forest arthropods. Pp. 678–686. In: D. J. Nelson and F. C. Evans (eds.). Symposium on Radioecology, CONF-670503. US AEC, Washington, DC.
Reichle, D. E., B. E. Dinger, N. T. Edwards, W. F. Harris and P. Sollins. 1973. Carbon flow and storage in a forest ecosystem. Pp. 345–365. In: G. E. Woodwell and E. V. Pecan (eds.). Carbon and the Biosphere. US AEC, CONF-720510, Washington, DC.
Reichle, D. E., R. V. O'Neill and W. F. Harris. 1975. Principles of energy and material exchange in ecosystems. Pp. 27–43. In: W. H. van Dobben and R. H. Lowe-McConnell (eds.). Unifying Concepts in Ecology. Dr. W. Junk B. V. Publ., The Hague.
Schurr, J. M. and M. N. Stamper. 1962. Model for the accumulation of strontium and calcium by recently molted crayfish (*Cambarus longulus longerostris* Ort.). Limnology and Oceanography 7:474–477.
Seastedt, T. R. and D. A. Crossley, Jr. 1984. The influence of arthropods on ecosystems. Bioscience 34:157–161.
Shanks, R. E. and J. S. Olson. 1961. First-year breakdown of leaf litter in southern Appalachian forests. Science 134:194–195.
Shugart, H. H. 1984. A Theory of Forest Dynamics: The Ecological Implications of Forest Succession Models. Springer-Verlag, New York, NY.
Shugart, H. H., R. A. Goldstein, R. V. O'Neill, and J. B. Mankin. 1974. TEEM: A terrestrial ecosystem energy model for forests. Oceol. Plant. 9:231–264.
Sollins, P., W. F. Harris and N. T. Edwards. 1976. Simulating the physiology of a temperate deciduous forest. Pp. 174–218. In: B. C. Patten (ed.). Systems Analysis and Simulation in Ecology, Volume 4. Academic Press, New York, NY.
Strandtmann, R. W. and D. A. Crossley, Jr. 1962. A new species of soil-inhabiting mite, *Hypoaspis marksi* (Acarina, Laelaptidae). Journal of the Kansas Entomological Society 35:180–185.
Tansley, A. G. 1935. The use and abuse of vegetational concepts and terms. Ecology 16:284–307.
Thomas, W. A. 1969. Accumulation and cycling of calcium by dogwood trees. Ecological Monographs 39:101–120.
Trabalka, J. R., J. A. Edmonds, J. M. Reilly, R. H. Gardner and D. E. Reichle. 1986. Atmospheric CO2 projections with globally averaged carbon cycle models. Pp. 534–560. In: J. R. Trabalka and D. E. Reichle (eds.). The Changing Carbon Cycle: A Global Analysis. Springer-Verlag, New York, NY.
Van Dyne, G. M. 1966. Ecosystems, systems ecology, and systems ecologists. ORNL-3957. Oak Ridge National Lab, Oak Ridge, TN. Also reprinted in: Patten, B. C. (ed.). 1995. Complex Ecology: The Part-Whole Relation in Ecosystems. Prentice Hall PTR, Englewood Cliffs, NJ. Pp. 1–27.

Van Hook, R. I., Jr. 1971. Energy and nutrient dynamics of spider and orthopteran populations in a grassland ecosystem. Ecological Monographs 41:1–26.

Waller, H. D. and J. S. Olson. 1967. Prompt transfers of cesium-137 to the soils of a tagged *Liriodendron* forest. Ecology 48:15–25.

Willard, W. K. 1960. Avian uptake of fission products from an area contaminated by low-level atomic wastes. Science 132:148–150.

Wiser, C. W. and D. J. Nelson. 1964. Uptake and elimination of cobalt-60 by crayfish. American Midland Naturalist 72:181–202.

Witherspoon, J. P., Jr. 1963. Cycling of Cesium-134 in white oak trees on sites of contrasting soil type and moisture, 1. 1960 growing season. Pp. 127–132. In: V. Schultz and A. W. Klement, Jr. (eds.). Radioecology. Reinhold Publ. Co., New York, NY.

Witherspoon, J. P. 1964. Cycling of Cesium-134 in white oak trees. Ecological Monographs 34:403–420.

Witherspoon, J. P. 1965. Radiation damage to forest surrounding an unshielded fast reactor. Health Physics 11:1637–1642.

Witherspoon, J. P. and G. N. Brown. 1965. Translocation of Cesium-137 from parent trees to seedlings of *Liriodendron tulipifera*. Botanical Gazette 126:181–185.

Witkamp, M. 1966. Decomposition of leaf litter in relation to environment, microflora, and microbial respiration. Ecology 47:194–201.

Witkamp, M. 1968. Accumulation of 137-Cs by *Trichoderma viride* relative to 137-Cs in soil organic matter and soil solution. Soil Science 106:309–311.

Witkamp, M. 1969a. Environmental effects on microbial turnover of some mineral elements, Part I—Abiotic factors. Soil Biology and Biochemistry 1:167–176.

Witkamp, M. 1969b. Environmental effects on microbial turnover of some mineral elements, Part II—Biotic factors. Soil Biology and Biochemistry 1:177–184.

Witkamp, M. 1969c. Cycles of temperature and carbon dioxide evolution from litter and soil. Ecology 50:922–924.

Witkamp, M. 1971a. Soils as components of ecosystems. Annual Review of Ecology and Systematics 2:85–110.

Witkamp, M. 1971b. Forest soil microflora and mineral cycling. Pp. 413–424. In: P. Duvigneaud (ed.). Productivity of Forest Ecosystems. UNESCO, Paris.

Witkamp, M. and B. Barzansky. 1968. Microbial immobilization of 137-Cs in forest litter. Oikos 19:392–395.

Witkamp, M. and M. L. Frank. 1964. First year of movement, distribution and availability of Cs137 in the forest floor under tagged tulip poplars. Radiation Botany 4:485–495.

Witkamp, M. and M. L. Frank. 1969. Cesium-137 kinetics in terrestrial microcosms. Pp. 635–643. In: D. J. Nelson and F. C. Evans (eds.). Symposium on Radioecology, CONF-670503. US AEC, Washington, DC.

Witkamp, M. and M. L. Frank. 1970. Effects of temperature, rainfall, and fauna on transfer of 137-Cs, K, Mg, and mass in consumer-decomposer microcosms. Ecology 51:465–474.

Witkamp, M., M. L. Frank and J. L. Shoopman. 1966. Accumulation and biota in a pioneer ecosystem of kudzu vine at Copperhill, Tennessee. Journal of Applied Ecology 3:383–391.

Witkamp, M. and J. S. Olson. 1963. Breakdown of confined and nonconfined oak litter. Oikos 14:138–147.

7

EL VERDE, PUERTO RICO

Introduction

The Radioecology project in the tropical rain forest at El Verde in the Luquillo Mountains of eastern Puerto Rico developed after most of the other major radioecology projects in the U.S. To some extent, by coming later the El Verde project benefited from earlier experiences, but the complexities of the rain forest environment generated new challenges and new discoveries on how ecosystems respond to radiation. The new aspect of operating a major Radioecology project in a rain forest was the point of the project. Woodwell and Houghton (1990) suggested that John Wolfe and the AEC wanted a tropical rain forest project to complement the other projects, and it may have been stimulated by developments in the idea for a sea-level canal excavated across Panama with the detonation of nuclear devices. However, the El Verde project seemed to emerge synergistically in Puerto Rico when connections were made between the Puerto Rico Nuclear Center (PRNC), the Institute of Tropical Forestry (ITF) and H. T. Odum, who happened by chance to be working part-time on applying new ecosystem thinking to the rain forest in the Luquillo Mountains.

Probably the key catalyst to the El Verde project was the existence of the PRNC. From the AEC's perspective, the PRNC offered an ideal support infrastructure for a major Radioecology project. Located in a suburb of San Juan, the PRNC was started in 1957 with AEC support, and at the beginning of the El Verde project it consisted of a large professional staff with a complete set of technical capabilities for conducting and supporting radiation research. Its association with the University of Puerto Rico was an additional strength of the PRNC. The long-term objective of the PRNC was to lead the development of atomic energy technology throughout Latin America, and having a major Radioecology project on the island supported this objective.

DOI: 10.4324/9781003284901-9

The Institute of Tropical Forestry of the U.S. Forest Service (now named the International Institute of Tropical Forestry) was important because it managed the land in the Luquillo Mountains and it employed experts who knew the ecology of the forest from a forestry perspective. The ITF, with headquarters in Rio Piedras near the PRNC headquarters, was an ideal partner to facilitate the radiation experiment in the rain forest. ITF granted access to the forest for the experiment and provided office space for project personnel. Frank Wadsworth, the director, and other ITF staff had extensively studied the tree species at El Verde and they had determined that the forest was an essentially undisturbed example of a Caribbean island tropical rain forest (Wadsworth 1970).

The final catalyst to the AEC funding of the rain forest project in the Luquillo Mountains was H. T. Odum, who had been coming to Puerto Rico for short research trips from his position as director of the Institute of Marine Science at the University of Texas. In the mid-1950s Odum had undertaken seminal studies on ecosystem structure and function, in the freshwater springs of Florida and in the coral reefs at Eniwetok in the Marshall Islands at the Pacific Proving Grounds with his brother. While at Texas on the Gulf of Mexico coast he continued this research approach and he began to formulate a general systems theory of ecosystems based on their metabolism. The research in Puerto Rico allowed him to extend his fieldwork on ecosystems to include tropical coastal systems (Golley et al. 1962, Odum et al. 1959) along with work in the rain forest of the Luquillo Mountains (Odum et al. 1963). Most of his previous work had involved aquatic ecosystems in which metabolism can be estimated fairly easily by measuring changes in dissolved oxygen over a diurnal cycle. Thus oxygen increases during the daylight period reflect net photosynthesis and decreases at night reflect community respiration; thus, once the data is corrected for physical diffusion, ecosystem metabolism can be calculated. In Puerto Rico for the first time he extended this methodological approach to terrestrial systems by measuring changes in carbon dioxide of tree leaves and soil plots with small, air-tight chambers. By replicating measures of CO_2 gas exchange and leaf mass over vertical prisms in the rain forest, Odum estimated the metabolism of the whole forest. This work in the rain forest allowed him to advance his energy theory of ecosystem dynamics based on the balance of primary production (P) and community respiration (R). He first articulated this theory in 1962, using data from the Puerto Rican rain forest at a conference on the suburban forest and ecology (Odum 1962). He states:

> The behavior of the montane forest seems similar to the microcosm of balanced aquaria in the constant temperature rooms at the Institute of Marine Science where Beyers (1962) has done exhaustive studies on the intricate coupling of P and R, one complementing the other in the supply of nutrients. . . . The unity of aquatic and terrestrial systems provides a powerful, single ecological theory.

Odum's research visits to Puerto Rico were short but intense in terms of the amount of fieldwork done, based on the notebooks in the Odum Library at the

University of Florida. Funding, essentially just for travel, came from a variety of sources. The key was initial support from the Rockefeller Foundation for trips in 1957 and 1958. During the first trips he made contact with Frank Wadsworth, who guided his work in the Luquillo Mountains. He was also assisted by personnel from the University of Puerto Rico, especially at their Marine Institute in Mayagüez. Odum made a critical contact with the recently established PRNC and eventually got funding from them for his final field trip in 1962. During this trip he made measurements of fallout radionuclides in the montane rain forest but also in the cloud forest at the highest elevations in the mountains. The findings of these first measurements of radioactivity in the rain forest were surprising and important to the initiation of the AEC-funded project to come:

> In October 1962, H. T. Odum, B. J. Copeland and R. Z. Brown, while on a biomass study at El Verde and in the Elfin Forest, collected epiphytic bromeliads and mosses on which preliminary measurements indicated fallout accumulations as high as those reported for arctic tundra. Shortly after this the U.S. Soil Conservation Service found fallout levels in the mountain forests sufficiently high to affect their field Geiger counters with readings much greater than normal background; they noted that readings increased from the ground to the crowns. The opportunity presented by a tagged forest was an important factor in the decision to start the entire project.
>
> *(Odum 1970a)*

Everything was perfectly aligned for funding of a large project by the AEC: background research on forest ecology and fallout radionuclides was available (Odum's contribution), a site on federal land with much forestry background information was available (ITF and Wadsworth's contribution) and a major atomic energy research and teaching facility with trained staff and modern lab facilities was nearby (PRNC's contribution). Odum had received research funds from the AEC for small projects on radioactivity while at the University of Texas Marine Institute, so he also had experience in Radioecology. Planning for the rain forest Radioecology project began in 1962 while Odum was still employed at the University of Texas. The initial grant proposal had Howard Teas, an ecologist working at the PRNC, as principal investigator with Odum listed as a consultant; however, Odum left Texas and became the principal investigator of the newly funded rain forest Radioecology project and an employee of the PRNC in early 1963.

John Bugher seems to have been a key player in orchestrating the initiation of the project. He was an influential scientist with the AEC (Warren 1971) and was the director of the PRNC from 1960 to 1966, which included the period that the rain forest project was conceived, planned and initiated. Earlier he had been director of the Division of Biology and Medicine of AEC, and he was the author of an early review paper on the environmental impacts of radionuclides (Bugher 1956). He was a member of the board of the Rockefeller Foundation, and it is interesting to speculate that he might have been involved in the initial funding of Odum's fieldwork in Puerto Rico in the late 1950s. Bugher became personally involved with measurements of gamma spectra of plant materials from El Verde; he coauthored a chapter (Kline et al. 1970a) and he wrote the foreword in the rain forest book (Odum and Pigeon 1970). Support of a scientist and administrator of Bugher's caliber certainly must have been influential in securing funding for the Radioecology project.

Site Description

The principal site of the AEC-supported Radioecology project was at the El Verde field station of U.S. Forest Service, at about 450 meters (1500 feet) elevation on the northwestern slope of the Luquillo Mountains in eastern Puerto Rico (Odum 1970b). This elevation placed the project in what is known as the Tabonuco forest type, named after the dominant tree species, the Tabonuco (*Dacryodes excelsa*). The original proposal for the rain forest Radioecology project (Teas and Odum 1962) lays out the rationale for locating the study at the El Verde location:

> The background of previous studies, the absence of a marked dry season, the relative ease of access, the proximity of supplies in San Juan, the cooperative support of resident scientists, and the absence of poisonous snakes make the El Verde area an ideal location for forest ecological research.

This site was under the control of the U.S. Forest Service, which granted access for the proposed work. This initial agreement and continuing support from Frank Wadsworth and the ITF were critical to the project's success.

The rain forest site, with an annual precipitation of 375 centimeters (148 inches) per year (Odum et al. 1970a), offered the AEC a distinct contrast to the sites of previous Radioecology research on the Pacific Ocean coral reefs, the Alaskan tundra, the Nevada desert and the temperate forests at several sites in the Eastern U.S. Because of the island location of the study site, the flora and fauna is reduced relative to what would be found at a continental site. Lawton (1995) described the Puerto Rican Tabonuco forest as "a simplified, depauperate version of mainland Caribbean rain forest", highlighting its biogeographic limitations. However, the following description from the original proposal (Teas and Odum 1962) conveys a richer impression of the Puerto Rican rain forest ecosystem:

> In the El Verde area tropical rain forest there is a full development of evergreen, broad-leafed foliage in several strata; a continuous excess of precipitation over evaporation; and other such characteristic features of the forest as tree ferns, bromeliads, lianas, buttress roots, numerous lizards, noisy tree frogs, tree top insects and birds, a thin mantle of organic soil and intermeshed surface roots. The trees are old and slow growing.

Experimental Design

The project was envisioned with three main objectives from the start (Odum 1964):

1 Measurement of effects of irradiation, direct and indirect, on the ecological system and its components.
2 Determination of the quantitative nature of the biogeochemical cycles within the forest through tracer and fallout studies so that the movement and fate of radionuclides may be predicted in rainforests.
3 Determination of the general structure and processes of the rainforest system as part of the general quest for principles that will permit an understanding of radiation effects and isotopic movements in any ecological system.

The work was scheduled in two phases, the first organized around a gamma source irradiation of the forest and the second organized around "biogeochemical studies of isotopes not directly related to the irradiation experiment but necessary to the understanding of forest processing of fallout."

The project began in May 1963 with H. T. Odum as the senior ecologist in the newly formed Terrestrial Ecology Program of the PRNC. Funding was granted from the Division of Biology and Medicine of the AEC under contract AT-(40–1)0–1835 with the PRNC of the University of Puerto Rico. The initial budget was for approximately $400,000 for 1963 and 1964, but the project would extend into the early 1970s and would total well more than $1,000,000.

The original proposal called for irradiation of two sites in the forest with carefully controlled sources of Cesium-137: a short duration test of 20 days and a long-term test of 2–5 years. It was also suggested that after several years an additional site "may be used for a short exposure of mixed radiation from a reactor". Ultimately a single study was settled on, with a 10,000-curie Cesium-137 source which irradiated the forest for 92 days, from January to April 1965. This constituted essentially an acute exposure, rather than the chronic exposure as was done in the temperate forest at Brookhaven, New York (Woodwell and Houghton 1990).

A circular layout of study plots, with a radius of 91 meters (300 feet), was used, with the cesium source in the middle of the irradiated plot, as had been done at Brookhaven for their radiation experiment (Woodwell 1963). The site with the cesium source was called the Radiation Center and a similar site located 160 meters (530 feet) away was called the South Control Center. Later, an additional plot of similar size and shape was established where the trees were cut to replicate the opening of the canopy due to tree mortality that occurred at the Radiation Center, but in this case without radiation effects. This plot was called the North Cut Center and its special use was to distinguish radiation impacts from impacts caused by opening the canopy.

The effects of irradiation were studied with three generic types of measurements, as described by McCormick (1970a):

> Radiation effects were estimated by: (1) comparing data collected before and after irradiation of the Radiation Center, (2) comparing data collected

> along a radiation-intensity gradient, and (3) comparing the Radiation and South Control Centers before and after irradiation.

A confounding factor in all studies was the trampling effect of researchers studying the centers, which started before irradiation began.

An ambitious work plan was proposed including systems-level work on metabolism and nutrient cycling. This plan covered research topics of a whole ecosystem study along with specific emphases at the organism and even cellular levels of organization that were appropriate for Radioecology. The presence of measurements of ecosystem function (forest metabolism and nutrient cycling) in the proposed research objectives was forward-looking for 1962. Most early studies of Radioecology focused on radiosensitivity of individual species or step-wise transfers along food chains rather than on systems processes. In a sense these early studies were done in ecosystems without being about ecosystems, but this changed in several of the AEC-funded projects of the early 1960s. The objective of biogeochemical nutrient cycling in the original proposal was probably of particular interest to the AEC due to the connection of fallout radionuclide retention:

> The recycling of high concentrations of nutrients with a lush ecosystem in spite of torrential rains is a primary function likely to be a sensitive indicator of forest function and the effects of radiation. Because a larger percentage of some nutrients ultimately come from the rain and because the forest has a very effective mechanism for preventing the loss of these nutrients faster than rates of influx, there are reasons for suspecting that fallout elements exist in higher concentrations relative to carrier elements than in most forests.
>
> *(Teas and Odum 1962)*

The inclusion of overall metabolism in the original proposal was an important goal for Odum and it turned into a signature effort of measuring metabolism by enclosing a portion of the whole forest for CO_2 input-output exchange. This particular study was really a small part of the overall rain forest project, but it received significant attention as a heroic effort of fieldwork.

The rain forest project was intended to be done in large part by scientists who would visit Puerto Rico for relatively short periods of intense research and then return to their home institutions. A number of these researchers were from other AEC Radioecology projects, and this approach was probably highly productive in terms of the sharing of methods and ideas across the many projects. Odum (1971) called this overall approach an "ecosystem task force" and it was an early example of teamwork of a large group of scientists that is an important part of modern field research. The actual full-time staff was quite small, including George Drewry, who served many functions and was resident at the field station during the project, local people with experience in the forest, including Alejo Estrada Pinto, and two graduate students: Ariel Lugo and Peter Murphy, who both got their Master of Science degrees on the project (Lugo 1965, see also Lugo 1970; Murphy 1967, see also Murphy 1970). Murphy was also a pilot and he provided flights over the forest for those needing aerial photography.

The Radiation Source and Safety

The cesium source was manufactured in Oak Ridge, Tennessee, at a company associated with the national lab. It was composed of a capsule of powdered cesium chloride that released gamma radiation during radioactive decay. The source was shipped to San Juan, loaded onto a pick-up truck and driven up the mountain (by a probably somewhat nervous Peter Murphy), then it was flown by helicopter to the concrete pad which had been built for it in the forest (Figure 7.1). A lead-lined "pig" or container enclosed the source with a covering (plug) that could be raised during radiation exposure or lowered for shielding. Electrical controls for the shielding apparatus were located at the El Verde Field Station and only H. T. Odum and George Drewry were authorized to raise the shielding plug over the source with coordination from the PRNC headquarters in Rio Piedras.

Fencing was erected around the source and signs were posted for safety warning of irradiation hazard. Two guards continuously patrolled the perimeter on eight-hour shifts to ensure no unauthorized people would enter the area during the three-month-long irradiation period. Details on the source, its operation, safety procedures and the radiation field generated by the source are provided in Section C of the rain forest book (Odum and Pigeon 1970). General comparisons with radiation sources used for irradiating ecosystems at other AEC Radioecology projects in New York, Georgia and Nevada were made by Odum and Drewry (1970). After the irradiation period the source was eventually removed from the forest and it was used again in the Enterprise Forest radioecology project in Wisconsin (Rudolph 1974).

Dosimetry

An important part of the experimental design was the use of dosimetry to record radiation dose. Dosimeters were deployed before the irradiation period began, and after the irradiation had concluded, they provided exposure values to match ecological data gathered by researchers. More than 1,000 small dosimeters were placed throughout the forest around the Radiation Center (see Table 1 in McCormick 1970a for the distribution). These included units placed on or in snails, lizards and rats.

The dosimetry effort also included a number of studies of radiation shielding by hard objects that absorbed the gamma radiation in the line of sight and thereby protected organisms behind them. In an example of this type of study, pairs of dosimeters were placed on 50 tree trunks in the Radiation Center, one on the side facing the radiation source and the other on the opposite side of the trunk, facing away from the source.

FIGURE 7.1 Radiocesium source in the rain forest at El Verde.

Source: From Odum and Drewry 1970.

Radiation Stress

On January 19, 1965, at 7:25 AM the shield was raised and the radiation experiment began. Although there were brief interruptions for scientific work and maintenance checks, the forest was exposed to the 10,000-curie Cs-137 source for 93 days, from January to April 1965. The direct damage from radiation occurred roughly in zones outward from the source. The most dramatic effect of near-complete mortality occurred at a radius of about 3 meters (10 feet), where the dosage was 100,000 r. Severe damage occurred within a radius of 10 meters (33 feet) from the source, with a dosage of about 10,000 r, and slight damage was noted as far as 30 meters (100 feet) from the source, where the dosage was about 4,000 r.

Different components of the forest ecosystem responded differently to the direct effects and to indirect or secondary effects caused by the irradiation. In terms of the vegetation, the overstory canopy trees responded more slowly than the understory plants. In part this was caused by the plug that was raised to expose the source and lowered to encase the source for safety. During the irradiation period the plug was raised and it thereby shielded the canopy directly above it from radiation damage. Because of the reduced dosage above the source, surrounding trees died back slowly. As described by Johnson and Atwood (1970), who were studying aerial photography at the site:

> At the end of the radiation period in late April 1965, evidence of only a small hole in the canopy could be inferred from the aerial photography. By April 1966, a year later, a somewhat large gap in the canopy had developed, approximately 12 by 18 m, whereas the affected zone on the ground was 40 m in diameter; by March 1967 the defoliated canopy approached 40 m in diameter.

The understory plants responded much faster to the radiation stress, with effects visible after one month. These effects were variable due to the dissected topography of the site and due to the localized shielding by rocks and tree trunks. Overall, the impacts of irradiation were dynamic and changing throughout the three-month exposure period, and they continued for more than two years after the source was encased and eventually removed from the site.

Some direct effects of radiation stress were clear and dramatic. For example, in his chapter on bryophytes Steere (1970) notes:

> In the radiation center, to a distance of perhaps 3 m from the source, rocks in the direct line of sight which were earlier covered with bryophytes were quite devoid of them. Whether the bryophytes were volatilized or crisped into dust by the high level [of] radiation, the fact remains that they have totally disappeared. Somewhat beyond the 3-m zone and in shielded spots within it, bryophytes in the direct line of sight, although apparently killed, were not destroyed completely, and the dead plants still remained attached to the rocks.

Impacts of irradiation on animals were variable between species and sometimes ambiguous. For examples, radiation effects were not detected for the dominant tree snail (Heatwole et al. 1970), litter microarthropods (Wiegert 1970) or rats (Weinbren et al. 1970) but increased mortality was found for anoles and tree frogs (Turner and

Gist 1970). Multiple examples of radiation stress were found for the arboreal, nest-building termites, which are important decomposers of dead wood in the rain forest (McMahan 1970). Nest damage and a reduction of occupancy for covered trails were found in the Radiation Center relative to the South Control Center. Also, a creative study of radiosensitivity was carried out in the laboratory with a Co-60 source at the PRNC. Artificial termite nests were constructed by placing fragments of occupied nests from the forest in plastic wastebaskets partially filled with damp leaves and twigs, which were then irradiated. Termites were found to be radiosensitive and it was speculated that social insects with complex behavior, such as termites, may be relatively more sensitive to radiation than other types of insects.

In general no effects were detected when microbes were studied, in part due to shielding by the leaf litter layer and the soil, but also because their populations turned over so rapidly that they were not capable of accumulating a large dosage during the 92 days of irradiation. Wiegert (1970) used this finding to explain the lack of statistical difference in litter decomposition rates between the Radiation and South Control Centers.

The most important indirect effect of irradiation was the gradual opening of the canopy, which subsequently changed the microclimate and light levels within the forest. These changes in the microenvironment of the rain forest were significant and they favored or affected species differently. The opening of the canopy at the Radiation Center acted in part like a light gap that occurs naturally as the result of tree falls caused by high winds. Comparative studies conducted in the North Cut Control site were intended to help differentiate radiation effects from effects due to the opening in the canopy. The study of lichens by Gannutz (1970) is a good example of the use of the comparisons with the North Cut Control Center. By studying lichen density in all three centers (the South Control, Radiation and North Cut Control) he was able to suggest that 25% of the total decline in lichens found in the Radiation Center might have been caused by indirect effects of changing microenvironment due to the opening of the canopy relative to direct radiation stress. As an aside, the chapter by Gannutz is also interesting because he had carried out studies of lichens at the Brookhaven radiation experiment (Woodwell and Gannutz 1967) and this experience allowed him to make direct comparisons of lichen response to radiation between the El Verde and Brookhaven radioecology projects.

To give an idea of the diversity of radiation impacts at El Verde, Table 7.1 lists 29 figures and tables showing examples of effects from throughout the rain forest book. In each case these examples illustrate radiation effects along one or more of the dimensions of the project experimental design. As synthesis of some of the main effects of the study, Figures 7.2 and 7.3 provide causal diagrams for the forest in the Radiation Center and the forest in the South Control Center, respectively. The main connected chains of causality run through tree growth in the diagrams. In the South Control Center, where the forest is essentially undisturbed, tree growth controls the internal environment, keeping the microclimate moist and cool and the level of light reaching the forest floor low. In the Radiation Center

TABLE 7.1 Examples of radiation stress in the rain forest at El Verde, Puerto Rico, that were reported in tables or figures. The page number for each Table or Figure in Odum and Pigeon (1970) is given.

Table or Figure	*Description of the Radiation Stress*	*Category of Effect*
Figure 11 P. C-31	location of the maximum leaf fall zone along the radiation field	direct effect
Figure 2 P. D-229	depression of lichen density along the radiation gradient	direct effect
Figure 1 P. G-23	increase in flowering of trees in the Radiation Center	direct effect
Figure 2 P. H-94	increase in leaf fall in the Radiation Center	direct effect
Figure 2 P. I-26	depression in Chlorophyll a in the Radiation Center	direct effect
Table 2 P. D-216	depression in bryophyte coverage in the Radiation Center	direct effect
Figure 72 P. D-62	increase in leaf fall over time in the Radiation Center	direct effect
Figure 73 P. D-62	map of tree mortality in the Radiation Center	direct effect
Figure 74 P. D-63	increase in gum streamers on trees in the Radiation Center	direct effect
Figure 9 P. D-281	increase in leaf fall in a tree species in the Radiation Center	direct effect
Figure 62 P. D-57	increase in yellow leaves in the Radiation Center	direct effect
Table 1 P. D-5	increase in fallen leaves in the Radiation Center	direct effect
Figure 18 P. D-170	tree mortality in the Radiation Center	direct effect
Figure 32 P. D-125	map of severe radiation damage in the Radiation Center	direct effect
Figure 26 P. D-122	vegetation profile of the Radiation Center showing tree damage	direct effect
Figure 34 P. D-126	general decrease in tree stem growth with radiation dose	direct effect
Figure 3 P. D-301	increase in bromeliad mortality with radiation dose	direct effect
Table 3 P. I-75	increase in herbivory as recorded by leaf holes in the Radiation Center	direct effect
Figure 15 P. E-117	increase of abandonment of termite tunnels in the Radiation Center	direct effect
Figure 9 P. E-64	depression in amphibian calls in the Radiation Center	direct effect
Table 3 P. E-30	reduction in anole abundance in the Radiation Center	direct effect
Table 1 P. G-22	increase in meiotic irregularities in *Palicourea riparia* after irradiation	genetic anomaly

(Continued)

TABLE 7.1 (Continued)

Table or Figure	*Description of the Radiation Stress*	*Category of Effect*
Figure 1 P. G-53	malformed testis of walkingstick insects after irradiation	genetic anomaly
Figure 16 P. D-118	increase in light intensity in the Radiation Center	obvious indirect effect
Figure 28 P. D-124	increase in air temperature in the Radiation Center	obvious indirect effect
Figure 25 P. D-97	increase in light as reflected in optical density in the Radiation Center	obvious indirect effect
Figure 8 P. D-209	depression of relative humidity in the Radiation Center	obvious indirect effect
Figure 11 P. B-75	depression in optical density in the Radiation Center	obvious indirect effect
Figure 2b P. D-289	increase in light levels in the Radiation Center	obvious indirect effect

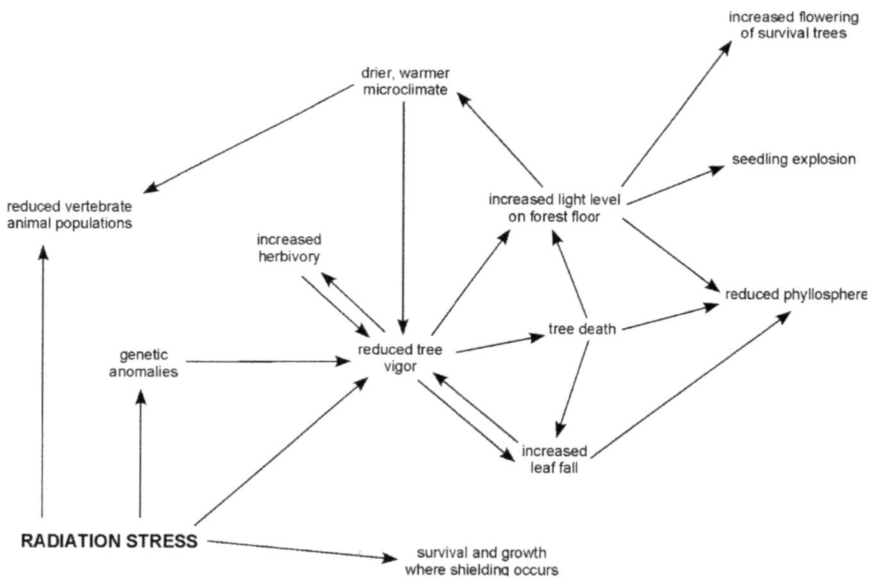

FIGURE 7.2 Causal diagram of the principal direct and indirect effects of irradiation on the rain forest at the Radiation Center.

the stress from radiation causes tree mortality that opens the canopy and changes the internal environment of the forest, leading to a cascade of indirect or secondary effects. Direct effects of radiation include generation of some scattered genetic anomalies, which are primarily covered in Section G of the rain forest book, and some reductions in at least vertebrate animal populations, partly from mortality

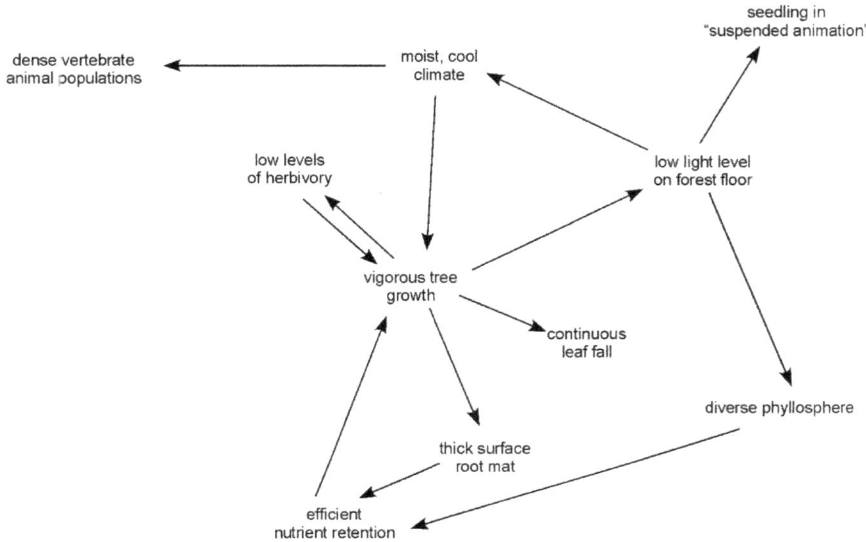

FIGURE 7.3 Causal diagram of elements of an undisturbed rain forest found at the South Control Center.

(such as lizards and frogs) and partly by movements away from the radiation source for more mobile organisms (such as birds).

Odum provides synthesis of aspects of the radiation project with his energy circuit diagrams in the Introduction to Section G (Figures 1 and 3 in Odum 1970d) and in his summary chapter, I-10 (Figure 22 in Odum 1970c). In each of these diagrams radiation is shown as an outside stressor that impacts the forest ecosystem. The focus of his two systems diagrams in the Introduction to Section G is on the effects of radiation at the cellular level of organization, while the diagram in Chapter I-10 relates to the mechanisms by which the forest recovers from radiation damage. These diagrams are models of whole ecosystems with different degrees of aggregation (see Figures 7.4 and 7.5 in following section). In a very different kind of synthesis, Odum provided several analogies of the appearance of the radiation stress in the rain forest. In Chapter D-1, which introduced the section on plants and the effects of radiation, he and his coauthors (Odum et al. 1970b) suggested that the inner zone of the Radiation Center looked like a "brush pile" with downed wood caused by tree mortality from radiation stress. In Chapter I-10, Odum (1970c) suggested that the inner zone of the Radiation Center resembled the forests of Appalachia after the chestnut blight had caused mortality of a dominant tree species in the mid-1900s. These analogies convey visual images of the damage caused by radiation in a manner similar to, but less abstract than, his energy circuit diagrams. Hecht and Cockburn (1990) provided additional imagery of the rain forest project with the following quote:

> For three months a patch of government-owned forest in El Verde, Puerto Rico, was bombarded with gamma rays. As the forest withered under this

> simulation of planetary holocaust, Odum led his team of young scientists to analyze the wounded ecosystem and from this study to deduce the structure and the functions of the forest.

Odum (1968) related several examples from the rain forest project in discussing an energy theory of stress in ecosystems. For example, he used the analogy of titration for the balance of radiation stress against ecosystem metabolism along the gradient extending from the cesium source into the rain forest. A breakpoint at 20 meters (65 feet) from the source was suggested as where stress balanced metabolism in comparable units of kilocalories. Alexander (1978) simulated an aggregated model of the rain forest along this radiation gradient, elaborating on Odum's original calculations, and Lugo (1978) extended this approach of comparing stress to metabolism for other types of ecosystems.

An interesting small-scale example of radiation stress was tested with microcosms in closed plastic dessicators that were essentially living models of the forest floor portion of the rain forest (Odum and Lugo 1970). Replicate microcosms were placed in the Radiation Center and the South Control Center before irradiation began. Microcosms in the Radiation Center received doses of at least 25,000 r over the three-month experimental period. After the experiment had concluded, metabolism was measured by gas exchange measurements of CO_2 concentration over diurnal curves. No difference in metabolism was found between the irradiated and the nonirradiated microcosms. The lack of effect was attributed to the dominance of microorganisms in the systems whose fast turnover may have conferred radioresistance to the irradiated microcosms. The microcosms developed with internal CO_2 concentrations that ranged over an order of magnitude, reflecting differences in the surviving biota inside. Because the microcosms were closed systems, the results were interpreted in regard to homeostasis of global atmospheric composition and the biota, to some extent anticipating the Gaia hypothesis (Lovelock 1979).

Radiation as a direct stressor on an individual organism affects DNA and cell function and a collection of papers at this level of organization is included in Section G of the rain forest book. This work collectively represented what Odum termed "field cytology", which is a unique contribution to the overall project. Odum's (1970d) own Introduction to Section G (entitled "Microscopic Order in the Forest") places this cellular-level research within the context of the overall project that was at the ecosystem level of organization. In this chapter, two aggregated energy circuit models are given showing cellular function and radiation stress. A literature review of cytological studies of radiation stress is given with an orientation around the reigning theory of biological radiosensitivity that had been advanced by A. H. Sparrow and associates at the Brookhaven Laboratory. Termed the target model, it holds that the amount of damage inflicted by irradiation is proportional to the exposed area of genetic material capable of absorbing the radiation energy. In practice the exposed area was taken to be the volume of the nucleus of the cell (e.g., nuclear volume) since this is the location of the genetic material. The model was used to predict relative radiosensitivity of a species with the relationship being the larger the nuclear volume of a species, the more sensitive it will be to radiation exposure. This model holds an important place in the field of Radioecology, and it was used and discussed in many of the AEC-funded projects. Through the association with the PRNC, and with consultation with Sparrow at Brookhaven, Odum was able to arrange to have the nuclear volumes of a number of rain forest plant species from El Verde measured by Koo and De Irizarry (1970), who were cell biologists, and these data were used in several chapters of the rain forest book to examine the target model as applied within the Radiation Center. For examples, Smith (1970) made comparisons of predicted radiosensitivities based on nuclear volume versus data on stem growth for four common tree species (Figures 35–38 in Chapter D-3); Watson (1970) compared nuclear volumes and branch elongation as a measure of stress for 21 woody plant species (Table 1 in Chapter D-16); and Steere (1970) compared nuclear volumes and survival of bryophytes. None of these studies supported the target model of species radiosensitivity and it was suggested that the tropical species of the rain forest were more radioresistant than was originally anticipated.

Odum (1970d) reviewed an alternative model for radiation resistance of a species related to the balance of damage and repair rates in cells that might better apply to rain forest species. He invited a chapter by Dienes (1970) from the Brookhaven Laboratory, who simulated dose-response graphs for both the target model and the damage-repair model. Although work on "field cytology" emphasized tests of the target model of radiosensitivity and the search for genetic anomalies in relation to radiation stress, the rain forest book contains other contributions at the cellular level of organization. Chlorophyll content of leaves, which was used as an index of photosynthetic capacity and as an indicator of radiation stress, was measured and reported in three chapters (Koo et al. 1970, Odum et al. 1970c, Odum and Cintron 1970). However, the most unique contribution at the cellular level of organization was Michael Canoy's (1970) estimation of total DNA content in the rain forest. He sampled DNA content of tree leaves and multiplied the average value by the leaf area index to estimate the mass of DNA per square meter of the forest. The purpose of this effort was to conceptualize the magnitude of information storage in the forest biota and to connect the cellular scale to the ecosystem. While this kind of thinking appears not to have been tried again by ecologists, one might wonder if it didn't anticipate later developments in environmental molecular biology, such as shotgun sequencing (Venter et al. 2004) and eDNA (Taberlet et al. 2012, Yaccoz 2012). Odum (1970d) concluded his introduction to the section on "field cytology" with the following quote which illustrates the multiple levels of organization represented in the rain forest book:

> The microscopic order is as much dependent on the homeostasis of the macroscopic organization of light fields, consumer control species, energy reserves for repair, and regenerative mechanisms as vice versa. The green

forest attains a quiet order at the macroscopic level even though it is a seething, selecting, noisy struggle with much error at the microscopic level.

Recovery and Self-Organization

On April 27, 1965, the irradiation period of the experiment ended as the cesium source was enclosed in the lead "pig" and later removed from the forest. After this time, researchers were able to freely enter the Radiation Center and take measurements of the stress that had occurred, as was summarized in the previous section, and to start studying the process of recovery and regrowth. Carl Jordan, who began work on the rain forest project in July 1966, later classified the irradiation experiment using criteria from ecological disturbance theory as being of moderate intensity, small size and short duration (Jordan 1985). These characteristics limited the relative impact of irradiation on the rain forest and fostered a relatively rapid recovery of the system after irradiation was stopped. Because the soil and litter layer was not physically disturbed by irradiation, nutrients were retained in the forest (Jordan and Kline 1972) which further facilitated recovery processes. These processes were even more rapid in the North Cut Control Center where the canopy was removed but no radiation stress occurred. Although canopy trees continued to die for a year in the Radiation Center after the source was capped, regrowth in the understory began immediately from seeds, from root sprouts and from surviving plants that had been shielded from irradiation by rocks, tree trunks or sloping land.

Odum used the medical metaphor of "healing" to describe the forest recovery (Odum 1970c, Odum et al. 1970b) and he included an energy circuit diagram of the mechanism in his summary chapter (Figure 7.4). In this model radiation is shown as a negative source of stress with the circle symbol to the right of the diagram (above the sources of positive energy from sunlight). The radiation turns some of the normal flora into damaged waste, which recycles through mineral cycling as fertilizing nutrients go back to interact with sunlight in primary production. The radiation also reduces shade (with mortality of the forest overstory), which in turn triggers germination of seeds of new plants and growth of repair species (such as *Cecropia* and *Palicourea*). The forests of the Luquillo Mountains are regularly exposed to natural disturbances such as hurricane and landslides (Scatena et al. 2012), and species such as *Cecropia peltata* are well adapted to the recovery process as pioneers in succession. However, the exposure to gamma irradiation created a new and different kind of disturbance, and the well-known pioneer species contributed less to forest recovery than they would have after a natural disturbance. Surprise occurred when *Palicourea riparia* emerged as the most dominant woody plant species in the early recovery of the forest in the Radiation Center. *P. riparia* was generally known as an abundant shrub/small tree in the understory of the rain forest (Little et al. 1974) and it was found to be common in surveys of the Radiation Center before irradiation (Desmarais and Helmuth 1970, Smith 1970). However, it was also found to have one of the largest nuclear volumes of all woody plants examined by Koo and De Irizarry (1970), which, according to the target theory, would have suggested that it would have been

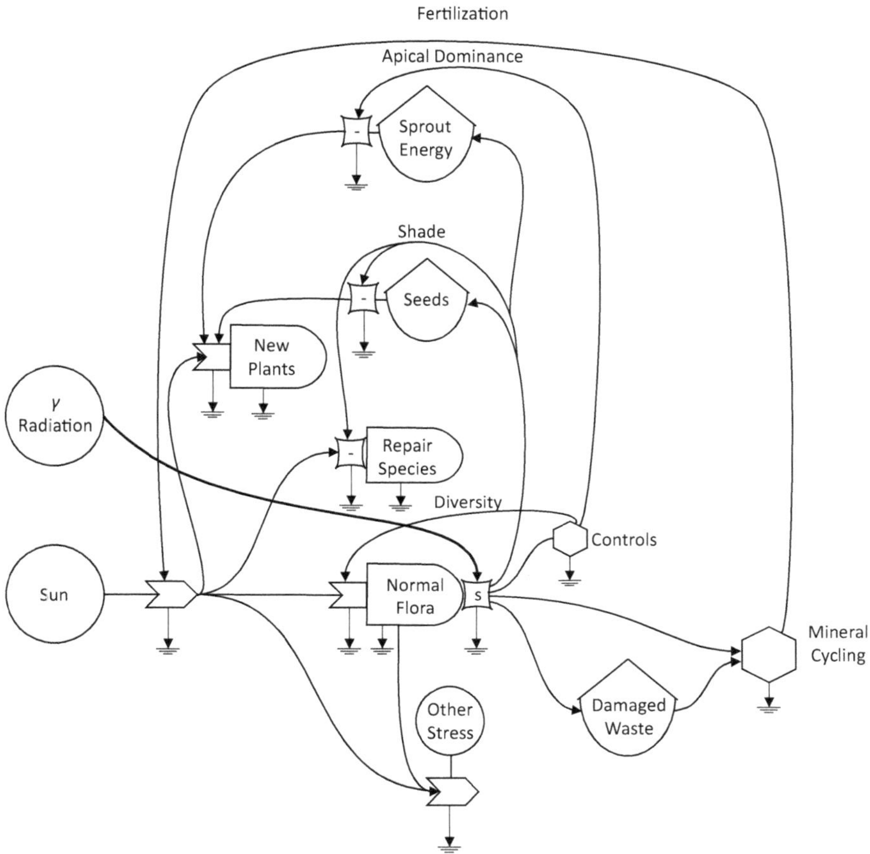

FIGURE 7.4 Energy circuit model of repair mechanisms that restore the forest after radiation stress. The normal flora is shown being replaced with new plants and repair species after radiation stress.

Source: From Odum 1970c.

relatively sensitive to radiation stress. Instead it was found to be the most radioresistant species, when seedlings of *P. riparia* increased dramatically during the irradiation experiment in the zone closest to the source in the Radiation Center (see, for example, Figure 6 in McCormick 1970b). This species was also found to be the most resistant woody plant in terms of branch growth (see Figure 3 in Watson 1970) and stem growth (see Figure 34 in Smith 1970) in the Radiation Center. These qualities of radioresistance were especially surprising because of its large nuclear volume, and this finding was an important stimulus to the search for more complicated mechanisms for understanding radiosensitivity of species in the rain forest.

Because *P. riparia* was abundant in the Radiation Center understory and because its flowering rate was high, it was fortunately chosen for cytological study before irradiation began (Venator and Koo 1970). These studies indicated that *Palicourea* normally had a high error rate in DNA replication and that irradiation increased this rate. The hypothesis was proposed that species more used to dealing with error normally, like *Palicourea*, would be the least affected by error-causing radiation. Thus, Odum (1970d) suggested that *P. riparia*'s ability to maintain itself with a high cellular error rate preadapted it for survival in the high exposure zone which, in turn, allowed it to respond quickly once the canopy was opened through defoliation by radiation stress and the light levels reaching the understory increased.

One of the long-term contributions of the discovery of *Palicourea*'s role in recovery of the rain forest was that it led to Odum's development of an abstract model for self-organization through loop reinforcement. This model is shown in the energy circuit language in Odum's introduction to the field cytology chapters (Figure 7.5). The model specifically describes gene maintenance and it represents the recursive process of evolution by natural selection. On the left-hand side of the diagram is the "choice generator", representing the propagules of an organism or genetic material at the cellular level, which are shown as replicate hexagons in the middle of the diagram. The choice generator is fueled by energy from outside the system, shown by the arrow entering the system from the left. The "chooser" is shown on the right-hand side of the diagram as a unit that "consumes the discards" of the ill-suited (e.g., less fit) propagules and indirectly selects the surviving propagules, shown at the top of the diagram, that loop back to the choice generator unit.

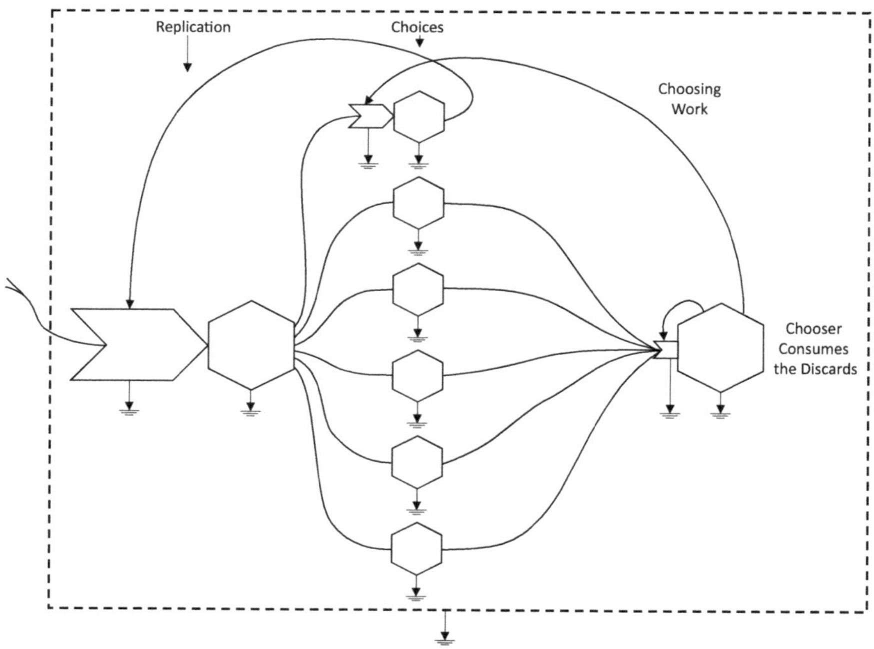

FIGURE 7.5 Energy circuit model of self-organization through loop reinforcement.
Source: Odum 1970d.

This occurs in gene replication at the cellular level or in reproduction in the life of an organism. Overall the model forms a loop that feeds back to reinforce the input of energy to the system through the creative process of Darwinian natural selection; in other words, it is an evolutionary algorithm. It is proposed here that the story of emergence of *P. riparia* from the species pool of available rain forest plants in the Radiation Center at El Verde inspired Odum's development of this general model through the context of cytological studies of the species genetic error rate. In this case the chooser was the radiation source which stressed individual plants. *Palicourea*'s ability to continually cycle through the loop of genetic replication with a high error rate pre-adapted it to survive in the zone of high radiation dose and to capture the increased light reaching the forest floor for photosynthesis while other species were killed by the radiation. The self-organizing capacity of the rain forest ecosystem is demonstrated by the loop reinforcement of a pre-adapted species (*P. riparia*) that thereby maximizes the energy flow of photosynthesis in the zone of highest radiation dosage. Odum generalized this model of self-organization in his book on energy systems (Odum 1971) and he continued to use and develop the model throughout his career.

Studies of the recovery of vegetation in the Radiation Center have continued since the irradiation period was concluded. This was an important responsibility of Carl Jordan when he joined the project, and he covered the early years of recovery (Jordan 1969, 1970a). Studies continued through the 1970s and 1980s (Silander 1985, Taylor et al. 1995), and most recently Lugo and Scalley (2014) provided a summary of biomass increase over time with a comparison with recovery from a hurricane for another forest plot in the Luquillo Mountains. This summary suggests biomass recovery at the Radiation Center is still somewhat less than half of what might be expected. Unfortunately, change over time for the South Control and the North Cut Control Centers appears not to have been studied for comparison with the Radiation Center. *P. riparia* has dominated the Radiation Center for sapling-sized trees over time as documented by Taylor et al. (1995). Subsequent studies in the Luquillo Mountains of recovery after other kinds of disturbances have documented a successional role for *P. riparia* (Lebron 1977, Walker et al. 2003, Aide et al. 1996, Crow 1980) but never as the pioneering dominant as occurred after irradiation at El Verde.

Phase II and the Sea Level Canal–Nuclear Excavation Project

Phase I of the rain forest project officially ended on June 30, 1966, a little more than a year after the irradiation experiment had ended. Odum left Puerto Rico for a faculty position at the University of North Carolina at this time and Jerry R. Kline became head of the rain forest project. Odum remained as a consultant and his contribution was to edit the book manuscript and work on an electrical analog model of the rain forest ecosystem.

As described in the original proposal, Phase II of the project moved away from the gamma irradiation experiment and covered new objectives. Recovery of the forest from irradiation remained a primary objective but the general focus changed to emphasize research on tracing mineral cycling function by studies with radionuclides, such as tritium, strontium and others. Much of the details of the Phase II research ultimately did not appear in the culminating rain forest book (Odum and Pigeon 1970), which mainly covered the Phase I irradiation experiment. Details of the Phase II research appear more fully in the annual reports of the rain forest project (Kline et al. 1967, Kline et al. 1968, Jordan and Drewry 1969, Clements et al. 1971) and in papers published in peer-reviewed journals.

Research started in Phase I on retention of fallout radionuclides was expanded in Phase II. Uptake and retention of fallout radionuclides in the forest was a significant finding of the project. Because of the high rainfall amounts, it might have been suspected that leaching of these elements might have occurred but the opposite was found. The consensus of multiple studies was that the community of epiphytic plants and other organisms on leaves and tree trunks was functioning to retain radionuclides scavenged from the air by dry deposition and from rainfall passing through the canopy. Odum (1970e) described the epiphytes in his introductory chapter to Section H, on mineral cycling in the rain forest book:

> A conspicuous property of the El Verde forest is the coating of epiphytic algae, mosses, liverworts, and microorganisms over the surfaces of most of the leaves, especially on the old slow-growing shade leaves. This epiphytic growth was found to hold fallout radioactivity and to play a prominent role in the mineral cycling of nutrients drawn up crown trees to be washed down with rain to drip upon the shade leaves where nutrients were apparently caught by the epiphytes.

Martin Witkamp (1970), a visiting microbial ecologist from the Oak Ridge National Laboratory, termed this community the phyllosphere and he demonstrated its role experimentally by applying a solution of radionuclides to leaves with and without epiphytes and then testing them for radioactivity. This experiment simulated sorption of minerals by foliage that would take place in the tree canopy during rain events. Higher retentions were found on the leaves covered with epiphytes than on clean leaves, demonstrating the role of the phyllosphere. Similar experiments reported in Kline et al. (1968) confirmed this work, and other chapters in Section H of the rain forest book provide supporting evidence about the role of the phyllosphere (Kline 1970, Odum et al. 1970d, Kline and Odum 1970). Furthermore, through applications of ^{134}Cs, ^{89}Sr and ^{54}Mn to plots of litter and soil in the forest, Kline and Mercado (1970) demonstrated that the radionuclides were not taken up by tree roots, which supported the role of epiphytes in determining the relatively high levels of radioactivity found on the leaves of the rain forest. Sollins and Drewry (1970) reported lower values of electrical conductivity in stemflow (i.e., rainfall moving down tree trunks) than in rainfall at the top of the canopy, which suggested uptake of minerals by epiphytes growing on the tree trunks. Finally, Tukey (1970) demonstrated the role of epiphytic bromeliads in mineral uptake.

All of this work makes a collective case for the importance of the phyllosphere in mineral and, in particular, radionuclide retention within the forest. Interestingly, typical rain forest adaptations for mineral retention such as surface root mats and mycorrhizae (symbiotic fungi attached to roots) were not found to be important in fallout radionuclide retention, though they were implicated in studies of the main plant nutrients of nitrogen and phosphorus (Edmisten 1970, Luse 1970).

Work in Phase II became directly related to the sea level canal program that was proposed by the AEC as part of its larger Project Plowshare, which sought peaceful uses of atomic energy. The proposal was to excavate a canal across the Isthmus of Panama at sea level by detonating nuclear devices. Since tropical rain forest occupied the proposed route for the canal, the results of research in the Puerto Rican rain forest project would have been indirectly relevant. The ecological studies directly connected to the Panama sea level canal proposal are an interesting part of the history of Radioecology and are mentioned in Chapter 9 of this book.

It is interesting to note that the sea level canal–nuclear excavation project was not mentioned in the original proposal for the El Verde rain forest project. However, it was explicitly mentioned at almost every opportunity thereafter as a justification for the project in Puerto Rico once it had been funded. For example, in a press release issued in Puerto Rico shortly before the irradiation experiment began, Odum (1965a) stated:

> If atomic explosions are to be used for the peaceful purposes of digging a new canal in Central America, what will be the consequences for the forests there? The experiment to be conducted at El Verde in January 1965, is designed to gain information pertinent to these questions.

Various divisions of the PRNC had been involved with research on the sea level canal–nuclear excavation project since at least 1961, so it would have been logical to justify the funding of the El Verde rain forest project with the sea level canal from the start. It is therefore curious that the sea level canal project was not mentioned in the original proposal. Baldivieso (2016), in a M.A. thesis, has suggested a number of unsubstantiated sinister conspiracy theories about Odum and the El Verde radiation project in connection with the sea level canal project. After the Phase I irradiation experiment at El Verde had been completed, Odum did receive new, direct support for research on the sea level canal proposal during the Phase II work: for a trip with Jerry Kline to measure fallout radionuclides in forests of Panama and in several other Central America locations (Kline and Odum 1970) and for the development of a model of hydrogen biogeochemistry in the rain forest that related to possible tritium movements that would be triggered by nuclear detonations (Odum and Bloom 1967, Odum et al. 1970e). However, the original funding for the rain forest project was not explicitly connected to the sea level–nuclear excavation proposal.

Carl Jordan, a recent Ph.D. from Rutgers University, joined the rain forest project at the start of Phase II with responsibilities to study the vegetation recovery in the Radiation Center, as mentioned earlier, along with mineral cycling research. Jordan developed an effective lysimeter for measuring water flows through the soil (Jordan 1968, 1970b) that was particularly useful in experimental studies in which radionuclides were injected into the rain forest and their movements were tracked. Kline, who had replaced Odum as director of the rain forest project, and Jordan embarked on an active program of study of element cycling using radioisotopes as tracers. The overall use of radioisotope tracers began early in the rain forest project with studies of P-32, but Phase II of the project emphasized tritium (Table 7.2).

Studies of tritium movements at El Verde were directly tied to the sea level canal program because this isotope of hydrogen is released in the detonation of thermonuclear devices, which were proposed to be used for excavation of the canal. The tritium released from the explosions would form radioactive water molecules that

TABLE 7.2 Listing of radionuclide injections during the El Verde radioecology project.

Date	*Injections*	*Data*
August 1963	two small trees tagged with ^{32}P	Figures 37 and 38, Chapter H-1
August 1963	^{32}P tagged in bromeliads	Table 3, Figures 3 and 4, Chapter H-11
August 1963	^{32}P tagged in three saplings	Table 1 and 2, Figure 1–3, Chapter H-12
August 1963	^{137}Cs, ^{32}P, ^{54}Mn, ^{89}Sr tagged on epiphyllic leaves	Table 1, Chapter H-14
January 1966	^{134}Cs, ^{85}Sr, ^{54}Mn added to soil	Figure 1, Chapter H-20
January 1966	^{15}N added to epiphyllae to test for N-fixation	Chapter H-21
August 1967	^{85}Sr, ^{134}Cs added to soil	Figures 2 and 3, Chapter H-19
Summer-Fall 1969	eight plants tagged with ^{32}P	Stark (1971)
Summer-Fall 1969	^{75}Se, ^{65}Zn, ^{59}Fe added to seedlings, fresh leaves and humus-root mat	Stark (1971)
Summer-Fall 1969	tritiated water added to litter overlying soil	Figure 2, Chapter H-8, see also Kline and Jordan (1968)
Summer-Fall 1969	three trees injected with tritiated water	Kline et al. (1970b)
Summer-Fall 1969	tritium added to "a plot of 3.7 square meters"	mentioned in Jordan and Drewry (1969)
Summer-Fall 1969	"multi-isotope experiment and a pulsed tritium experiment"	mentioned in Jordan and Drewry (1969)
Summer-Fall 1969	"tritium uptake, residence half time . . . in a secondary successional tropical rain forest"	mentioned in Jordan and Drewry (1969)
Summer-Fall 1969	". . . injecting a tree with tritium . . ."	mentioned in Jordan and Drewry (1969)

would enter the normal water cycle and become embedded into the forest ecosystem. The need to contribute to a general understanding of tritium dynamics was thus a strong motivation to study it in Phase II at El Verde. At least six experimental injections of tritium into the rain forest were made with additions to the soil surface and directly into plants (Jordan and Drewry 1969). Some of this research involved personnel with experience from the Nevada Test Site, where tritium dynamics had been traced in the past from bomb testing. Several papers on the El Verde tritium work were published in journals (Kline and Jordan 1968, Kline et al. 1970b, Jordan et al. 1970). Three chapters on tritium research were included in the rain forest book (H-6, H-7, H-8), including a description of the initial tracer work that had been published in the journal *Science* (Kline and Jordan 1970). Chapters H-6 (Odum et al. 1970e) and H-7 (Bloom and Raines 1970) were based on a subcontract to Odum from the Battelle Memorial Institute, which managed the sea level canal program for the AEC (Odum and Bloom 1967). Chapter H-6 is a detailed description of essentially the biogeochemistry of hydrogen in the El Verde rain forest with an input-output budget that was used to calibrate analog and digital computer models. Simulation of these models followed a unit pulse of tritiated water through the rain forest as would occur with a thermonuclear device explosion used for excavation.

As an aside, starting in the late 1970s several surveys of residual radioactivity from the injections of radioisotopes were made (Anonymous 1993). It was determined that all but one of the injections had decayed enough that no exposure risk remained. The exception was an individual tree located about 700 meters east of the El Verde Field Station that had been injected with 0.46 millicuries of Cs-137 in 1968. Because some of the cesium had moved from the tree to the surrounding soil, about 43 kilograms of contaminated soil was excavated and removed from around the base of the tree and taken to Oak Ridge, Tennessee, for disposal. Residual activity was still above background levels in the tree and the adjacent soil in a survey done in 1993, but it was determined that the area around the tree was sufficiently isolated from potential human exposure and that the activity level was low enough that no more action needed to be taken.

A final part of Phase II of the project was the "giant cylinder" erected 40 meters (130 feet) east of the Radiation Center to measure metabolism and evapotranspiration of the rain forest (Odum and Jordan 1970). This was literally a giant cylinder of plastic hung from towers that enclosed a prism of rain forest 20 meters (67 feet) high and 18 meters (60 feet) wide with a surface area of 213 square meters (2,300 square feet). Air was drawn down through the cylinder by a large fan (1.8 meters [6 feet] in diameter) attached to the plastic wall at the bottom of the cylinder. With appropriate conversion and calibration factors, metabolism was measured by the difference in CO_2 content, and evapotranspiration was measured by the difference in humidity, between air entering the cylinder at the top and air leaving at the bottom by the fan. Basically, the systems and procedures used were the same as those used for monitoring metabolism and evapotranspiration of single leaves in small chambers (Odum et al. 1970f).

The giant cylinder was part of the original proposal for the rain forest project (Teas and Odum 1962). It seems clear that Odum had wanted to use this approach for measuring metabolism since the Rockefeller project in the late 1950s when he and his collaborators calculated metabolism from summation of small chamber measurements (Odum et al. 1963). The work was started in 1964 but it was interrupted by the irradiation experiment. The work was resumed in the summer of 1965 and continued into 1966 and 1967.

A monumental effort of physical work was required to acquire, to construct and to carry up the mountain at El Verde all the equipment needed to build and operate the giant cylinder. The towers were made of aluminum supported with steel wire and placed on concrete pads. The large fan was operated by a 15-horsepower gasoline engine. The plastic could be manually raised (when measurements were made) or lowered (between measurements) with a system of lines and pulleys. The first plastic curtain used was "a cheap grade of 10-mil polyethylene" which had to be sewn together with needle and thread and it proved not to be adequate; a new type of plastic was ultimately settled on in 1966 that was "thinner, stronger, lighter" and could be bonded with adhesives (Odum and Jordan 1970).

It is not clear if the giant cylinder was intended to contribute to the original irradiation experiment in Phase I of the project, but usable measurements were not made with it until 1966, long after irradiation had ended. Ultimately only six full measurement runs of metabolism and evapotranspiration were made over periods of one to three days. Averages of 0.47 g $C/m^2/hr$ for respiration and 1291 g water/ m^2/day for evapotranspiration were reported in Table 4 of Chapter I-9 (Odum and Jordan 1970). As was anticipated beforehand, many problems with these measurements occurred that limited their accuracy (see, for example, Ordway 1969). However, the heroic nature of the effort to enclose a portion of a whole forest in plastic for measuring metabolism was mentioned often in the years to come; for example, Jordan, who had the main responsibility to operate the giant cylinder after Odum left the project, called it "a spectacular subproject" of the entire study (Jordan 2001), and Mitsch and Day (2004) included the giant cylinder as an example of Odum "thinking big" in ecosystem studies.

At the end of the chapter about the giant cylinder, Odum and Jordan (1970) made suggestions for improving the technique, but the new designs were never tested. In Odum's large Cypress project in Florida during the early 1970s, he and his graduate students used a smaller-scale approach for measuring metabolism and transpiration of a forested wetland (Brown et al. 1984).

Of significance, work on the giant cylinder became directly associated with the sea level canal–nuclear excavation project assessment in Panama with additional funding from the Battelle Memorial Institute after Odum left Puerto Rico in 1966. Water budget studies at the giant cylinder site were used in constructing models of tritium movements. In fact, Odum and Jordan (1970) began their chapter on the giant cylinder with the following sentence, which justifies the work in relation to the proposed sea level canal project:

> Determining the total carbon metabolism and the total evapotranspiration of water is essential to the understanding and management of rain

forests, including such isotope problems as predicting the fate of tritium that might be released in excavation for an interocean canal with atomic fusion detonations.

Heritage

The heritage of the AEC-supported rain forest project at El Verde was, first and foremost, from a historical perspective, the "rain forest book" itself (Odum and Pigeon 1970). Even though it emphasized Phase I of the project, the rain forest book encapsulated the ecosystem-level approach taken by Odum and supported by the AEC administrators (primarily Wolfe and Bugher) who funded the project.

Because of its significance within the scientific community at the time, the rain forest book was formally reviewed in eight major journals: *Journal of Ecology* (Ashton 1972), *Journal of Applied Ecology* (Bartley 1973), *Bioscience* (Blair 1971), *Science* (Connell 1971), *Journal of Animal Ecology* (Elton 1972), *Quarterly Review of Biology* (Holt 1972), *Nature* (Richards 1971) and *Ecology* (Kannowski 1971). This was and remains an unprecedented amount of formal review for a book. The people selected by the journal editors to review the book were also noteworthy, including the famous English ecologists Charles Elton and P. W. Richards. The book reviews were generally positive, as has been the case with others who have commented on the book (Ewel 2003, Hecht and Cockburn 1990, Jacobs 1981). The historian Joel Hagen (1992) was critical of the book and the project as a whole, but he seems to have misread the ecological literature about it (for example, in citing only two of the eight formal reviews of the book). Coleman (2010) praised the rain forest project as a model for "how a scientific task force might go about studying a complex ecosystem" but found the project book "diffuse and disjointed". He goes on to suggest these opposing qualities also characterized the "big ecology" of the IBP projects. Finally, Raby (2019) provided a social scientist's perspective on the rain forest project as a cultural form of land use within Puerto Rico.

At the time of publication there was really no other text like the rain forest book in terms of the breadth of coverage of a single ecosystem. It has been suggested that the Cape Thompson text (Wilimovsky and Wolfe 1966), which was also from an AEC-funded project (see Chapter 5), was comparable, but it lacked some of the elements found in the rain forest book. To at least some extent there does seem to be a continuity between these big-scale, AEC-funded projects, both backed by John Wolfe, the head of AEC's environmental programs. The Cape Thompson project brought together a team approach in one spot to describe an ecosystem facing impact by an atomic energy development, but it lacked a unifying systems theme like that which Odum brought to the rain forest project. It is claimed here that the progression seen from the Cape Thompson project to the El Verde project represents one dimension of the evolution of the ecosystem concept, fostered by the AEC, which is the overarching theme of the present book.

One of the interesting things about reviews of the rain forest book is the almost uniform reference to its mass, which was relatively large for a conventional book. Of course this is unusual since mass is seldom a quality used in assessing a book. The rain forest book is large: 4.9 kilograms (10.8 pounds). The large size of both the Cape Thompson book and the rain forest book was possible because the publisher was the AEC, a not-for-profit entity from a publishing perspective. It is unlikely that any commercial publisher would have marketed such large texts.

In a sense the rain forest book is a physical manifestation of the ecosystem it describes, which is also large, complicated and complex. The holism of the ecosystem—the need to think of it as whole system rather than a simple collection of parts—was reinforced by the editorial decision of insisting on publishing all of the project reports in one book. This holistic sense would not have occurred if all of the chapters had been published separately at different times and in different journals, as was done for example in the Brookhaven AEC project.

A special feature of the project and the book is the synthesis of the parts into descriptions of the whole ecosystem. This synthesis takes multiple forms. At the simplest level, the creation of biomass tables and pyramids shows relative contributions of different components in the forest to the whole system biomass. Measurements of metabolism are also used to show relative contributions of components and the estimates from the giant cylinder show total system metabolism. The overall values of forest structure and function derived from the El Verde project were some of the first ever made at the ecosystem scale for a forest and, after more than 40 years, the data continue to be used for comparison with more recent studies of other rain forests (see, for example, Table 10.3 in Ghazoul and Sheil 2010).

The introductory chapters to each section of the book and the final summary chapter, written by Odum, tie together the sections and add literature review and perspective for the studies done in the rain forest. The summary chapter is particularly noteworthy, with 98 pages, 48 figures and 47 tables (Odum 1970c). Subsections in the summary chapter organize it around important topics of a modern ecosystem study and include Forest Biomass, Energy Flows, Mineral Cycles, Species Diversity, Radiation Effects and Suggestions for Future Studies.

A final important method of synthesis is the use of diagrammatic and computer simulation models that are found throughout the book and in several later publications. Most of the models are portrayed in the energy circuit language, which is a symbolic modeling language that Odum developed (see Figures 7.4 and 7.5). He explains the use of the energy circuit symbols early in the rain forest book, and he uses system diagrams composed of the symbols in a number of chapters. These diagrams show outside forcing functions and internal components and flow pathways. Interactions between components and feedback circuits representing control networks are embedded in the diagrams. Aggregation is used to lump together components and processes in order to keep the models simple enough to visually describe system structure. For some readers, however, these models are not effective in communicating a systems perspective. For example, in his review of the rain forest book, Joseph Connell "found Odum's circuit diagrams difficult to understand" (Connell 1971). However, for other readers Odum's energy circuit diagrams are an important synthesis of system structure and function (Brown 2004).

An early version of a system model of forest energy flow was used in the project proposal. Odum used this old modeling style in some of his early, classic systems papers (Odum 1956, 1957, 1960), but he developed his formal energy circuit language during the 1960s, in fact especially during the rain forest project, and he used it through the rest of his career.

There are 65 individual systems diagrams in 11 chapters (10% of the total) of the rain forest book and some of the most important are given in the summary chapter (Odum 1970c). The book also contains five simulation models: three using an analog computer and two using a digital computer. Three of the models simulate the movement of a pulse of tritium through the forest (see Chapters H-6 and H-7) as related to the nuclear excavation proposal for a sea level canal across Panama, while the other two simulate aspects of forest metabolism (see Chapters I-3 and I-10). These computer models were some of the first simulated at the ecosystem scale, and their use supports Kwa's (1993) contention that the need for understanding atomic energy impacts stimulated the origin of the field of Systems Ecology. The rain forest book also contains passive analog circuits as models of the forest ecosystem in Chapters I-3 and I-10. This was an early form of dynamic modeling, using standard electrical components, that Odum favored before analog and digital computers were available to ecologists (Kangas 2004). Odum (1965b) designed a complicated passive electrical analog model of the rain forest that was intended as a research tool and as an education exhibit but it seems never to have been built. Overall, the use of three forms of hardware for simulation (passive analogs, operational analog computers and digital computer) places the rain forest book at a crossroads in the history of Systems Ecology and it demonstrates Odum's interest in dynamic modeling of ecosystems (Kangas 1995). In addition to the computer simulation models contained in the rain forest book, Jordan and Kline (Jordan et al. 1973, Jordan and Kline 1976; see also Jordan et al. 1972, Kline 1973) constructed and simulated a model of cycling of fallout radionuclides showing long-term dynamics, using data from their fieldwork at El Verde. It would be interesting to go back and measure strontium content in the forest now to test their predictions.

Phase II work in the rain forest project continued through the late 1960s. Jerry Kline left the project in 1968 and Carl Jordan left in 1969, both for the Argonne National Laboratory, in Illinois, where they continued to work in Radioecology. Jordan later moved to the University of Georgia and became an international authority on rain forest mineral cycling and ecology (see, for example, Jordan 1985). By 1972 there is no more mention of the original rain forest project in the annual reports of the PRNC, but the Terrestrial Ecology group continued research on new objectives. Eventually the group became part of the Center for Energy and Environment Research (CEER) with joint support from the new Department of

Energy and the University of Puerto Rico. Forestry research continued through the U.S. Forest Service's ITF, later renamed the International Institute of Tropical Forestry (Brown et al. 1983). In 1988 the entire Caribbean National Forest of the Luquillo Mountains became a part of the National Science Foundation–supported Long Term Ecological Research (LTER) network with research at all scales focusing on disturbance ecology, especially in terms of hurricane impacts to the forest. The Luquillo LTER is built on the foundation of research carried out in the AEC's rain forest project in the 1960s, at least to some extent. Much new research is filling in some of the missing elements from the original project, such as study of the control circuits that regulate ecosystem metabolism (see, for example, Dial and Roughgarden 1995, Beard et al. 2003). Three major edited volumes of ecological research (Reagan and Waide 1996, Brokaw et al. 2012, Gonzalez et al. 2013) have been published on the continuing work at El Verde and the greater Luquillo Mountains which, along with a massive number of publications referenced on the LTER's website, make the rain forest of the AEC's radioecology project one of the most widely studied ecosystems on the planet.

Odum finished editing the rain forest book while at the University of North Carolina. The book was published in1970 with the help of Robert Pigeon, a technical editor at the Oak Ridge National Laboratory. Odum left soon thereafter for a faculty position at the University of Florida, where he stayed until the end of his career. At Florida the scope of his research continued to expand, from field ecology to general energy systems theory and environmental accounting, but he continued to have connections to the Puerto Rican rain forest by serving as an advisor to CEER and to the Luquillo LTER. He and his colleagues applied systems models and his unique form of energy analysis to rain forest issues at multiple scales (Odum 1995, Scatena et al. 2002). Some of this work was highly abstract, such as his calculation of the energy embodied in the DNA of the rain forest (Odum 1995, 1996), based on work done in the original project by Michael Canoy 30 years earlier. Ariel Lugo, who did his Master of Science research under Odum on the AEC rain forest project (Lugo 1970) and later rose to be director of the IITF, has written several informative essays about Odum's contributions to rain forest ecology during and after the radiation project at El Verde (Lugo 1995, 2004, 2016).

Literature Cited

Aide, T. M., J. K. Zimmerman, M. Rosario and H. Marcano. 1996. Forest recovery in abandoned cattle pastures along an elevational gradient in northeastern Puerto Rico. Biotropica 28:537–548.

Alexander, J. F. 1978. Energy basis of disasters and the cycles of order and disorder. PhD Dissertation, University of Florida, Gainesville, FL.

Anonymous. 1993. Radiological measurements, Study Area 4, El Verde Research Station, Luquillo Forest, Luquillo, Puerto Rico. Contract DE-AC05–76OR00033 from the U.S. Department of Energy, Energy/Environment Systems Division, Oak Ridge Associated Universities, Oak Ridge, TN.

Ashton, P. S. 1972. Review of: A tropical rain forest. Journal of Ecology 60:623–625.

Baldivieso, C., Jr. 2016. Irradiating eden: The El Verde experiment and the atomic energy commission's Nuclear Prospecting in Latin America, 1954–1970. M. A. Thesis, University of Maryland, Baltimore County.

Bartley, D. D. 1973. Review of: A tropical rain forest. Journal of Applied Ecology 10:355–356.

Beard, K. H., A. K. Eschtruth, K. A. Vogt, D. J. Vogt and F. N. Scatena. 2003. The effects of the frog *Eleutherodactylus coqui* on invertebrates and ecosystem processes at two scales in the Luquillo Experimental Forest, Puerto Rico. Journal of Tropical Ecology 19:607–617.

Beyers, R. 1962. The metabolism of twelve aquatic laboratory micro-ecosystems. PhD Dissertation, University of Texas, Austin, TX.

Blair, W. F. 1971. Rain forest ecology, review of: A tropical rain forest. Bioscience 21:925–926.

Bloom, S. G. and G. E. Raines. 1970. Kinetic model study of hydrogen flow through the El Verde forest system. Pp. H-123–H-128. In: H. T. Odum and R. F. Pigeon (eds.). A tropical rain forest. USAEC Technical Information Center, Oak Ridge, TN.

Brokaw, N., T. A. Crowl, A. E. Lugo, W. H. McDowell, F. N. Scatena, R. B. Waide and M. R. Willig (eds.). 2012. A Caribbean Forest Tapestry. Oxford University Press, Oxford, UK.

Brown, M. T. 2004. A picture is worth a thousand words: Energy systems language and simulation. Ecological Modelling 178:83–100.

Brown, S. L., S. W. Cowles and H. T. Odum. 1984. Metabolism and transpiration of cypress domes in north-central Florida. Pp. 145–163. In: K. C. Ewel and H. T. Odum (eds.). Cypress Swamps. University Presses of Florida, University of Florida, Gainesville, FL.

Brown, S. L., A. E. Lugo, S. Silander and L. Liegel. 1983. Research History and Opportunities in the Luquillo Experimental Forest. U.S. Department of Agriculture Forest Service, General Technical Report SO-44. Southern Forest Exp. Stn., New Orleans, LA.

Bugher, J. C. 1956. Effects of fission material on air, soil, and living species. Pp. 831–850. In: W. L. Thomas, Jr. and Others (eds.). Man's Role in Changing the Face of the Earth. University of Chicago Press, Chicago, IL.

Canoy, M. J. 1970. Deoxyribonucleic acid in rain forest leaves. Pp. G-69–G-70. In: H. T. Odum and R. F. Pigeon (eds.). A Tropical Rain Forest. USAEC Technical Information Center, Oak Ridge, TN.

Clements, R. G., G. E. Drewry and R. J. Lavigne. 1971. The Rain Forest Project Annual Report. PRNC-147, Rio Piedras, PR.

Coleman, D. C. 2010. Big Ecology: The Emergence of Ecosystem Science. University of California Press, Berkeley, CA.

Connell, J. H. 1971. Project at El Verde, review of: A tropical rain forest. Science 172:831–832.

Crow, T. R. 1980. A rainforest chronicle: A 30-year record of change in structure and composition at El Verde, Puerto Rico. Biotropica 12:42–55.

Desmarais, A. P. and B. T. Helmuth. 1970. Effects of 137Cs radiation on vegetation and optical density at El Verde. Pp. D-77–D-103. In: H. T. Odum and R. F. Pigeon (eds.). A Tropical Rain Forest. USAEC Technical Information Center, Oak Ridge, TN.

Dial, R. and J. Roughgarden. 1995. Experimental removal of insectivores from rain forest canopy: Direct and indirect effects. Ecology 76:1821–1834.

Dienes, G. J. 1970. A kinetic model of radiation response in field irradiation. Pp. C-49–C-52. In: H. T. Odum and R. F. Pigeon (eds.). A Tropical Rain Forest. USAEC Technical Information Center, Oak Ridge, TN.

Edmisten, J. 1970. Preliminary studies of the nitrogen budget of a tropical rain forest. Pp. H-211–H-215. In: H. T. Odum and R. F. Pigeon (eds.). A Tropical Rain Forest. USAEC Technical Information Center, Oak Ridge, TN.

Elton, C. 1972. Review of: A tropical rain forest. Journal of Animal Ecology 41:250–252.
Ewel, J. J. 2003. Howard Thomas Odum (1924–2002). Bulletin of the Ecological Society of America 84:13–15.
Gannutz, T. P. 1970. A study of the lichens of an irradiated rain forest. Pp. D-227–D-233. In: H. T. Odum and R. F. Pigeon (eds.). A Tropical Rain Forest. USAEC Technical Information Center, Oak Ridge, TN.
Ghazoul, J. and D. Sheil. 2010. Tropical Rain Forest Ecology, Diversity, and Conservation. Oxford University Press, Oxford, UK.
Golley, F., H. T. Odum and R. F. Wilson. 1962. The structure and metabolism of a Puerto Rican red mangrove forest in May. Ecology 43:9–19.
Gonzalez, G., M. R. Willig and R. B. Waide (eds.). 2013. Ecological Gradient Analyses in a Tropical Landscape. Ecological Bulletins 54. John Wiley and Sons, Oxford, UK.
Hagen, J. B. 1992. An Entangled Bank: The Origins of Ecosystem Ecology. Rutgers University Press, New Brunswick, NJ.
Heatwole, H., A. Rossy, I. Colorado and R. Amadeo. 1970. Effects of radiation on a population of the Puerto Rican tree snail, *Caracolus caracolla*. Pp. E-17–E-24. In: H. T. Odum and R. F. Pigeon (eds.). A Tropical Rain Forest. USAEC Technical Information Center, Oak Ridge, TN.
Hecht, S. and A. Cockburn. 1990. The Fate of the Forest. Harper Perennial Publ., New York, NY.
Holt, B. R. 1972. Review of: A tropical rain forest. Quarterly Review of Biology 47:113–114.
Jacobs, M. 1981. The tropical rain forest: A First Encounter. Springer-Verlag, Berlin.
Johnson, P. L. and D. M. Atwood. 1970. Aerial sensing and photographic study of the El Verde rain forest. Pp. B-63–B-78. In: H. T. Odum and R. F. Pigeon (eds.). A Tropical Rain Forest. USAEC Technical Information Center, Oak Ridge, TN.
Jordan, C. F. 1968. A simple, tension-free lysimeter. Soil Sci. 105:81–86.
Jordan, C. F. 1969. Recovery of a tropical rain forest after gamma irradiation. Pp. 88–98. In: D. J. Nelson and F. C. Evans (eds.). Symposium on Radioecology, Proceedings of the Second National Symposium. CONF-670503, U.S. Atomic Energy Commission, Washington, DC.
Jordan, C. F. 1970a. Vegetative sprouting following irradiation of a tropical rain forest. Pp. D-305–D-308. In: H. T. Odum and R. F. Pigeon (eds.). A Tropical Rain Forest. USAEC Technical Information Center, Oak Ridge, TN.
Jordan, C. F. 1970b. Movement of ^{85}Sr and ^{134}Cs by the soil water of a tropical rain forest. Pp. H-201–H-204. In: H. T. Odum and R. F. Pigeon (eds.). A Tropical Rain Forest. USAEC Technical Information Center, Oak Ridge, TN.
Jordan, C. F. 1985. Nutrient Cycling in Tropical Forest Ecosystems. John Wiley & Sons, Chichester.
Jordan, C. F. 2001. Tropical ecology with the Institute of Ecology. In: G. W. Barrett and T. L. Barrett (eds.). Holistic Science: The Evolution of the Georgia Institute of Ecology (1940–2000). Taylor & Francis Publishers, New York, NY.
Jordan, C. F. and G. E. Drewry. 1969. The Rain Forest Project Annual Report. PRNC-129, Rio Piedras, PR.
Jordan, C. F. and J. R. Kline. 1972. Mineral cycling: Some basic concepts and their application in a tropical rain forest. Annual Reviews of Ecology and Systematics 3:33–50.
Jordan, C. F. and J. R. Kline. 1976. Strontium-90 in a tropical rain forest: 12th-yr validation of a 32-yr prediction. Health Physics 30:199–201.
Jordan, C. F., J. R. Kline and D. S. Sasscer. 1972. Relative stability of mineral cycles in forest ecosystems. American Naturalist 106:237–253.

Jordan, C. F., J. R. Kline and D. S. Sasscer. 1973. A simple model of strontium and manganese dynamics in a tropical rain forest. Health Physics 24:477–489.

Jordan, C. F., J. J. Koranda, J. R. Kline and J. R. Martin. 1970. Tritium movement in a tropical ecosystem. Bioscience 20:807–812.

Kangas, P. 1995. Contributions of H. T. Odum to ecosystem simulation modeling. Pp. 11–18. In: C. A. S. Hall (ed.). Maximum Power: The Ideas and Applications of H. T. Odum. University Press of Colorado, Niwot, CO.

Kangas, P. 2004. The role of passive electrical analogs in H. T. Odum's system thinking. Ecological Modelling 178:101–106.

Kannowski, P. B. 1971. Tropical radioecology, Review of: A Tropical Rain Forest. Ecology 52:1131–1132.

Kline, J. R. 1970. Retention of fallout radionuclides by tropical forest vegetation. Pp. H-191–H-197. In: H. T. Odum and R. F. Pigeon (eds.). A Tropical Rain Forest. USAEC Technical Information Center, Oak Ridge, TN.

Kline, J. R. 1973. Mathematical simulation of soil-plant relationships and soil genesis. Soil Science 115:240–249.

Kline, J. R. and C. F. Jordan. 1968. Tritium movement in soil of tropical rain forest. Science 160:550–551.

Kline, J. R. and C. F. Jordan. 1970. Tritium movement in soil a tropical rain forest. Pp. H-129–H-131. In: H. T. Odum and R. F. Pigeon (eds.). A Tropical Rain Forest. USAEC Technical Information Center, Oak Ridge, TN.

Kline, J. R., C. F. Jordan and G. E. Drewry. 1967. The Rain Forest Project Annual Report. PRNC-103, Rio Piedras, PR.

Kline, J. R., C. F. Jordan and G. E. Drewry. 1968. The Rain Forest Project Annual Report. PRNC-119, Rio Piedras, PR.

Kline, J. R., J. R. Martin, C. F. Jordan and J. J. Koranda. 1970b. Measurement of transpiration in tropical forests with tritiated water. Ecology 51:1068–1073.

Kline, J. R. and N. Mercado. 1970. Preliminary studies of radionuclide cycling in understory plants in the rain forest. Pp. H-205–H-209. In: H. T. Odum and R. F. Pigeon (eds.). A Tropical Rain Forest. USAEC Technical Information Center, Oak Ridge, TN.

Kline, J. R. and H. T. Odum. 1970. Comparisons of the amounts of fallout radionuclides in tropical forests. Pp. H-181–H-186. In: H. T. Odum and R. F. Pigeon (eds.). A Tropical Rain Forest. USAEC Technical Information Center, Oak Ridge, TN.

Kline, J. R., H. T. Odum and J. C. Bugher. 1970a. Effect of gamma radiation on leaching of 137Cs and 54Mn from tropical forest foliage and litter. Pp. H-187–H-190. In: H. T. Odum and R. F. Pigeon (eds.). A Tropical Rain Forest. USAEC Technical Information Center, Oak Ridge, TN.

Koo, F. K. S. and E. R. de Irizarry. 1970. Nuclear volume and radiosensitivity of plant species at El Verde. Pp. G-15–G-20. In: H. T. Odum and R. F. Pigeon (eds.). A Tropical Rain Forest. USAEC Technical Information Center, Oak Ridge, TN.

Koo, F. K. S., H. T. Odum and E. R. de Irizarry. 1970. Effect of gamma radiation on chlorophyll a content in bromeliads. Pp. D-295–D-298. In: H. T. Odum and R. F. Pigeon (eds.). A Tropical Rain Forest. USAEC Technical Information Center, Oak Ridge, TN.

Kwa, C. 1993. Radiation ecology, systems ecology and the management of the environment. Pp. 213–249. In: M. Shortland (ed.). Science and Nature, Essays in the History of the Environmental Sciences. British Society of the History of Science, Oxford, UK.

Lawton, J. H. 1995. Tabonuco trees, trophic manipulations and Tarzan: Lizards in a tropical forest canopy. TREE 10:392–393.

Lebron, M. L. 1977. An autecological study of *Palicourea riparia* bentham (Rubiaceae): An ecologically important species in the recovery of a disturbed tropical rain forest in Puerto Rico. Ph. D. Dissertation, University of North Carolina, Chapel Hill, NC.

Little, E. L., Jr., R. O. Woodbury and F. H. Wadsworth. 1974. Trees of Puerto Rico and the Virgin Islands, Second Volume. Agriculture Handbook No. 449. U.S. Department of Agriculture, Washington, DC.
Lovelock, J. E. 1979. Gaia: A New Look at Life on Earth. Oxford University Press, Oxford, UK.
Lugo, A. E. 1965. Photosynthetic studies on rain forest seedlings of *Cecropia peltata*, *Acanthocephalus cadamba*, *Dacryodes excelsa* and *Sloanea berteriana*. M. S. Thesis, University of Puerto Rico, Rio Piedras, PR.
Lugo, A. E. 1970. Photosynthetic studies on four species of rain forest seedlings. Pp. I-81–I-102. In: H. T. Odum and R. F. Pigeon (eds.). A Tropical Rain Forest. USAEC Technical Information Center, Oak Ridge, TN.
Lugo, A. E. 1978. Stress and Ecosystems. Pp. 62–101. In: J. H. Thorp and J. W. Gibbons (eds.). Energy and Environmental Stress in Aquatic Systems. U.S. Department of Energy CONF-771114. NTIS, Springfield, VA.
Lugo, A. E. 1995. Contributions of H. T. Odum to Tropical Ecology. Pp. 23–24. In: C. A. S. Hall (ed.). Maximum Power: The Ideas and Applications of H. T. Odum. University Press of Colorado, Niwot, CO.
Lugo, A. E. 2004. H. T. Odum and the Luquillo experimental forest. Ecological Modelling 178:65–74.
Lugo, A. E. 2016. A glimpse of the Tropics through Odum's macroscope. Pp. 251–258. In: M. R. Willig and L. R. Walker (eds.). Long-Term Ecological Research, Changing the Nature of Scientists. Oxford University Press, Oxford, UK.
Lugo, A. E. and T. H. Scalley. 2014. Research in the Luquillo Experimental Forest has advanced understanding of tropical forests and resolved management issues. Pp. 435–468. In: D. C. Hayes, S. L. Stout, R. H. Crawford and A. P. Hoover (eds.). USDA Forest Service Experimental Forests and Ranges, Research for the Long Term. Springer, New York, NY.
Luse, R. A. 1970. The phosphorus cycle in a tropical rain forest. Pp. H-161–H-166. In: H. T. Odum and R. F. Pigeon (eds.). A Tropical Rain Forest. USAEC Technical Information Center, Oak Ridge, TN.
McCormick, J. F. 1970a. Patterns of radiation exposure in the tropical rain forest. Pp. C-41–C-48. In: H. T. Odum and R. F. Pigeon (eds.). A Tropical Rain Forest. USAEC Technical Information Center, Oak Ridge, TN.
McCormick, J. F. 1970b. Direct and indirect effects of gamma radiation on seedling diversity and abundance in a tropical forest. Pp. D-201–D-212. In: H. T. Odum and R. F. Pigeon (eds.). A Tropical Rain Forest. USAEC Technical Information Center, Oak Ridge, TN.
McMahan, E. A. 1970. Radiation and the termites at El Verde. Pp. E-105–E-122. In: Odum and R. F. Pigeon (eds.). A Tropical Rain Forest. USAEC Technical Information Center, Oak Ridge, TN.
Mitsch, W. J. and J. W. Day, Jr. 2004. Thinking big with whole-ecosystem studies and ecosystem restoration: A legacy of H. T. Odum. Ecological Modelling 178:133–155.
Murphy, P. G. 1967. Tree growth at El Verde and the effects of ionizing radiation. M. S. Thesis, Syracuse University, Syracuse, NY.
Murphy, P. G. 1970. Tree growth at El Verde and the effects of ionizing radiation. Pp. D-141–D-173. In: H. T. Odum and R. F. Pigeon (eds.). A Tropical Rain Forest. USAEC Technical Information Center, Oak Ridge, TN.
Odum, H. T. 1956. Primary production in flowing water. Limnology and Oceanography 1:102–117.
Odum, H. T. 1957. Trophic structure and productivity of Silver Springs, Florida. Ecological Monographs 27:55–112.

Odum, H. T. 1960. Ecological potential and analogue circuits for the ecosystem. American Scientist 48:1–8.
Odum, H. T. 1962. Ecological tools and their use, man and the ecosystem. Pp. 57–75. In: P. E. Waggoner and J. D. Ovington (eds.). Proceedings of the Lockwood Conference on the Suburban Forest and Ecology, Bulletin 652. Connecticut Agricultural Experiment Station, New Haven, CT.
Odum, H. T. 1964. The Rain Forest Project, Work in Progress, Scientific Results, and Proposals. PRNC-34. Puerto Rico Nuclear Center, Rio Piedras, PR.
Odum, H. T. 1965a. Information on the PRNC Terrestrial Ecology Program. Press Release for "Members of the Press, Radio and Television", January 8, 1965, San Juan, PR. (seen in the Odum Library, University of Florida, Gainesville, FL).
Odum, H. T. 1965b. An Electrical Network Model of the Rain Forest Ecological System. PRNC-67. Puerto Rico Nuclear Center, Rio Piedras, PR.
Odum, H. T. 1968. Work circuits and systems stress. Pp. 81–138. In: H. E. Young (ed.). Symposium on Primary Productivity and Mineral Cycling in Natural Ecosystems. College of Life Sciences and Agriculture, University of Maine, Orono, ME.
Odum, H. T. 1970a. The AEC rain forest program. Pp. C-3 to C-21. In: H. T. Odum and R. F. Pigeon (eds.). A Tropical Rain Forest. USAEC Technical Information Center, Oak Ridge, TN.
Odum, H. T. 1970b. The El Verde study area and the rain forest systems of Puerto Rico. Pp. B-3–B-32. In: H. T. Odum and R. F. Pigeon (eds.). A Tropical Rain Forest. USAEC Technical Information Center, Oak Ridge, TN.
Odum, H. T. 1970c. Summary: An emerging view of the ecological system at El Verde. Pp. I-191–I-289. In: H. T. Odum and R. F. Pigeon (eds.). A Tropical Rain Forest. USAEC Technical Information Center, Oak Ridge, TN.
Odum, H. T. 1970d. Introduction to Section G, Microscopic order in the forest. Pp. G-3–G-14. In: H. T. Odum and R. F. Pigeon (eds.). A Tropical Rain Forest. USAEC Technical Information Center, Oak Ridge, TN.
Odum, H. T. 1970e. Rain forest structure and mineral-cycling homeostasis. Pp. H-3–H-52. In: H. T. Odum and R. F. Pigeon (eds.). A Tropical Rain Forest. USAEC Technical Information Center, Oak Ridge, TN.
Odum, H. T. 1971. Environment, Power, and Society. John Wiley & Sons, New York, NY.
Odum, H. T. 1995. Tropical forest systems and the human economy. Pp. 343–393. In: A. E. Lugo and C. Lowe (eds.). Tropical Forest Management and Ecology. Springer-Verlag, New York, NY.
Odum, H. T. 1996. Environmental Accounting. John Wiley & Sons, New York, NY.
Odum, H. T. with the assistance of J. Ruiz-Reyes. 1970. Holes in leaves and the grazing control mechanism. Pp. I-69–I-80. In: H. T. Odum and R. F. Pigeon (eds.). A Tropical Rain Forest. USAEC Technical Information Center, Oak Ridge, TN.
Odum, H. T., W. Abbott, R. K. Selander, F. B. Golley and R. F. Wilson. 1970c. Estimates of chlorophyll and biomass of the Tabonuco Forest of Puerto Rico. Pp. I-3–I-20. In: H. T. Odum and R. F. Pigeon (eds.). A Tropical Rain Forest. USAEC Technical Information Center, Oak Ridge, TN.
Odum, H. T. and S. G. Bloom. 1967. Hydrogen budget and compartments in the rain forest at El Verde, Puerto Rico, pertinent to consideration of tritium metabolism, with Appendix, Mathematical formulation of the hydrogen budget model. Prepared under Purchase Order No. S-3946, U.S. A. E. C. prime Contract No. AT(26–1)-171. Report to the Battelle Memorial Institute, Columbus, OH.

Odum, H. T., G. A. Briscoe and C. B. Briscoe. 1970d. Fallout radioactivity and epiphytes. Pp. H-167–H-176. In: H. T. Odum and R. F. Pigeon (eds.). A Tropical Rain Forest. USAEC Technical Information Center, Oak Ridge, TN.
Odum, H. T., P. R. Burkholder and J. Rivero. 1959. Measurements of productivity of turtle grass flats, reefs, and the Bahia Fosforescente of southern Puerto Rico. Publications of the Institute of Marine Science, University of Texas 6:159–170.
Odum, H. T. and G. Cintron. 1970. Forest chlorophyll and radiation. Pp. I-21–I-34. In: H. T. Odum and R. F. Pigeon (eds.). A Tropical Rain Forest. USAEC Technical Information Center, Oak Ridge, TN.
Odum, H. T., B. J. Copeland and R. Z. Brown. 1963. Direct and optical assay of leaf mass of the lower montane rain forest of Puerto Rico. Proceedings of the National Academy of Sciences, U.S. 49:429–434.
Odum, H. T. and G. Drewry. 1970. The cesium source at El Verde. Pp. C-23–C-36. In: H. T. Odum and R. F. Pigeon (eds.). A Tropical Rain Forest. USAEC Technical Information Center, Oak Ridge, TN.
Odum, H. T., G. Drewry and J. R. Kline. 1970a. Climate at El Verde, 1963–1966. Pp. B-347–B-418. In: H. T. Odum and R. F. Pigeon (eds.). A Tropical Rain Forest. USAEC Technical Information Center, Oak Ridge, TN.
Odum, H. T. and C. F. Jordan. 1970. Metabolism and evapotranspiration of the lower forest in a giant plastic cylinder. Pp. I-165–I-190. In: H. T. Odum and R. F. Pigeon (eds.). A Tropical Rain Forest. USAEC Technical Information Center, Oak Ridge, TN.
Odum, H. T. and A. Lugo. 1970. Metabolism of forest-floor microcosms. Pp. I-35–I-56. In: H. T. Odum and R. F. Pigeon (eds.). A Tropical Rain Forest. USAEC Technical Information Center, Oak Ridge, TN.
Odum, H. T., A. Lugo, G. Cintron and C. F. Jordan. 1970f. Metabolism and evapotranspiration of some rain forest plants and soil. Pp. I-103–I-164. In: H. T. Odum and R. F. Pigeon (eds.). A Tropical Rain Forest. USAEC Technical Information Center, Oak Ridge, TN.
Odum, H. T., A. M. Moore and L. A. Burns. 1970e. Hydrogen budget and compartments in the rain forest. Pp. H-105–H-122. In: H. T. Odum and R. F. Pigeon (eds.). A Tropical Rain Forest. USAEC Technical Information Center, Oak Ridge, TN.
Odum, H. T., P. Murphy, G. Drewry, F. McCormick, C. Schinhan, E. Morales and J. A. McIntyre. 1970b. Effects of gamma radiation on the forest at El Verde. Pp. D-3–D-76. In: H. T. Odum and R. F. Pigeon (eds.). Tropical Rain Forest. USAEC Technical Information Center, Oak Ridge, TN.
Odum, H. T. and R. F. Pigeon (eds.). 1970. A Tropical Rain Forest. USAEC Technical Information Center, Oak Ridge, TN.
Ordway, D. F. 1969. An aerodynamicist's analysis of the Odum cylinder approach to net CO2 exchange. Photosynthetica 3:199–209.
Raby, M. 2019. "Slash-and-burn ecology": Field science as land use. History of Science 57:441–468.
Reagan, D. P. and R. B. Waide (eds.). 1996. The Food Web of a Tropical Rain Forest. University of Chicago Press, Chicago, IL.
Richards, P. W. 1971. Forest radioecology. Nature 234:233–234.
Rudolph, T. D. 1974. The Enterprise Radiation Forest research facility. Pp. 1–16. In: T. D. Rudolph (ed.). Enterprise Radiation Forest, Preirradiation Ecological Studies. TID-26113, Technical Information Center, Office of Information Services, U.S. Atomic Energy Commission, Washington, DC.
Scatena, F. N., J. F. Blanco, K. H. Beard, R. B. Waide, A. E. Lugo, N. Brokaw, W. L. Silver, B. L. Haines and J. K. Zimmerman. 2012. Disturbance Regime. Pp. 164–200. In: N. Brokaw and six others (eds.). A Caribbean Forest Tapestry. Oxford University Press, Oxford, UK.

Scatena, F. N., S. J. Doherty, H. T. Odum and P. Kharecha. 2002. An EMERGY Evaluation of Puerto Rico and the Luquillo Experimental Forest. Gen. Tech. Rep. IITF-GTR-9. U.S. Department of Agriculture, Forest Service. International Institute of Tropical Forestry. Rio Piedras, PR.

Silander, S. 1985. Succession at the El Verde Radiation Site, a 17 Year Record. CEER-T-205, Center for Energy and Environment Research, U.S. Department of Energy, Rio Piedras, PR.

Smith, R. F. 1970. The vegetation structure of a Puerto Rican rain forest before and after short-term gamma irradiation. Pp. D-103–D-140. In: H. T. Odum and R. F. Pigeon (eds.). A Tropical Rain Forest. USAEC Technical Information Center, Oak Ridge, TN.

Sollins, P. and G. Drewry. 1970. Electrical conductivity and flow rate of water through the forest canopy. Pp. H.-137–H-153. In: H. T. Odum and R. F. Pigeon (eds.). A Tropical Rain Forest. USAEC Technical Information Center, Oak Ridge, TN.

Stark, N. 1971. Radiotracer studies of nutrient cycles. Pp. 130–142. In: R. G. Clements, G. E. Drewry and R. J. Lavigne (eds.). The Rain Forest Project Annual Report. PRNC-147, Rio Piedras, PR.

Steere, W. C. 1970. Bryophyte studies on the irradiated and control sites in the rain forest at El Verde. Pp. D-213–D-226. In: H. T. Odum and R. F. Pigeon (eds.). A Tropical Rain Forest. USAEC Technical Information Center, Oak Ridge, TN.

Taberlet, P., E. Coissac, M. Hajibabaei and L. H. Rieseberg. 2012. Environmental DNA. Molecular Ecology 21:1789–1793.

Taylor, C. M., S. Silander, R. B. Waide and W. J. Pfeiffer. 1995. Recovery of a tropical forest after gamma irradiation: A 23-year chronicle. Pp. 258–285. In: A. E. Lugo and C. Lowe (eds.). Tropical Forests: Management and Ecology. Springer-Verlag, New York, NY.

Teas, H. and H. T. Odum. 1962. Radioecology of a Tropical Rain Forest at El Verde, Puerto Rico. A research proposal to the Division of Biology and Medicine, U.S. Atomic Energy Commission. Washington, DC.

Tukey, H. B., Jr. 1970. Leaching of metabolites from foliage and its implication in the tropical rain forest. Pp. H-155–H-160. In: H. T. Odum and R. F. Pigeon (eds.). A Tropical Rain Forest. USAEC Technical Information Center, Oak Ridge, TN.

Turner, F. B. and C. S. Gist. 1970. Observations of lizards and tree frogs in an irradiated Puerto Rican forest. Pp. E-25–E-50. In: H. T. Odum and R. F. Pigeon (eds.). A Tropical Rain Forest. USAEC Technical Information Center, Oak Ridge, TN.

Venator, R. and F. K. S. Koo. 1970. Inherent and radio-induced cytological abnormalities in *Palicourea riparia*. Pp. G-21–G-38. In: H. T. Odum and R. F. Pigeon (eds.). A Tropical Rain Forest. USAEC Technical Information Center, Oak Ridge, TN.

Venter, J. C. and 22 co-authors. 2004. Environmental genome shotgun sequencing of the Sargasso Sea. Science 304:66–74.

Wadsworth, F. H. 1970. Review of past research in the Luquillo Mountains of Puerto Rico. Pp. B-33–B-46. In: H. T. Odum and R. F. Pigeon (eds.). A Tropical Rain Forest. USAEC Technical Information Center, Oak Ridge, TN.

Walker, L. R., D. J. Lodge, S. M. Guzman-Grajales, and N. Fetcher. 2003. Species-specific seedling responses to hurricane disturbance in a Puerto Rican rain forest. Biotropica 35:472–485.

Warren, S. 1971. Obituary, John Clifford Bugher. Radiation Research 46:412–413.

Watson, H. 1970. Branch elongation and the radiation field. Pp. D-287–D-294. In: H. T. Odum and R. F. Pigeon (eds.). A Tropical Rain Forest. USAEC Technical Information Center, Oak Ridge, TN.

Weinbren, M. P., B. M. Weinbren, W. B. Jackson and J. B. Villella. 1970. Studies on the roof rat (*Rattus rattus*) in the El Verde forest. Pp. E-169–E-181. In: H. T. Odum and R. F. Pigeon (eds.). A Tropical Rain Forest. USAEC Technical Information Center, Oak Ridge, TN.

Wiegert, R. G. 1970. Effects of ionizing radiation on leaf fall, decomposition, and the litter microarthropods of a montane rain forest. Pp. E-89–E-100. In: H. T. Odum and R. F. Pigeon (eds.). A Tropical Rain Forest. USAEC Technical Information Center, Oak Ridge, TN.

Wilimovsky, N. J. and J. W. Wolfe (eds.). 1966. Environment of the Cape Thompson Region, Alaska. U.S. Atomic Energy Commission, Division of Technical Information, Washington, DC.

Witkamp, M. 1970. Mineral retention by epiphyllic organisms. Pp. H-177–H-179. In: H. T. Odum and R. F. Pigeon (eds.). A Tropical Rain Forest. USAEC Technical Information Center, Oak Ridge, TN.

Woodwell, G. M. 1963. Design of the Brookhaven experiment on the effects of ionizing radiation on a terrestrial ecosystem. Radiation Botany 3:125–133.

Woodwell, G. M. and T. P. Gannutz. 1967. Effects of chronic gamma radiation on lichen communities of a forest. American Journal of Botany 54:1210–1215.

Woodwell, G. M. and R. A. Houghton. 1990. The experimental impoverishment of natural communities: Effects of ionizing radiation on plant communities, 1961–1976. Pp. 9–24. In: G. M. Woodwell (ed.). The Earth in Transition. Cambridge University Press, Cambridge, UK.

Yaccoz, N. G. 2012. The future of environmental DNA in ecology. Molecular Ecology 21:2031–2038.

8

THE BROOKHAVEN NATIONAL LABORATORY

Introduction

The Brookhaven National Laboratory (BNL) was the site of the longest-running experiment in Radioecology and what became one of the best-known studies of the effects on ionizing radiation on ecosystems. The BNL, located on Long Island, New York (about 100 kilometers [60 miles] east of New York City), was one of the first national labs established by the newly formed Atomic Energy Commission (AEC) in 1947. The AEC sought to establish national labs throughout the United States and the BNL was chosen for the northeastern region of the country. The lab was established at the U.S. Army's former Camp Upton and the site is 2150 hectares (5300 acres) in area. The BNL had the first nuclear reactor to be constructed in the U.S. after World War II, along with particle accelerators and other facilities for atomic energy research. A range of non–weapons research on atomic energy was and continues to be carried out at the BNL, with a strong emphasis on medicine and biology. An example of the lab's role in radiobiology is that its staff was designated to study the effects of radiation exposure of the inhabitants of the Marshall Islands who were accidentally contaminated by the radioactive fallout from the Bravo test on March 1, 1954 (Cohn et al. 1960).

Arnold Sparrow and the Radiosensitivity of Plants

One of the best-known biological researchers at Brookhaven was Arnold H. Sparrow. He worked at the BNL from its opening in 1947 through to his death in 1976 (Conger 1977, Van't Hof 1977). With AEC support, Sparrow conducted extensive research of the effects of radiation on plants. He formed a group of colleagues and technicians that conducted research in the laboratory, in greenhouses and in the field, studying numerous higher plant species. The responses of plants to radiation that

DOI: 10.4324/9781003284901-10

were studied included "four main types: chromosome breakage, somatic mutation, growth inhibition and lethality" (Sparrow et al. 1961). Of note, he developed large-scale studies in a cultivated field at the BNL where a Cobalt-60 source was placed in the corner of the experimental plot to produce gamma irradiation (Sparrow 1960). Plants were grown in concentric circles (or arcs) around this source, thereby establishing experimental gradients of exposure outward from the central source. The Co-60 source was placed in an upright metal tube mounted on a cylindrical lead shield or "pig". The source could be elevated to emit radiation or lowered for shielding by means of an underground cable attached to a windlass in a control building located a safe distance away from the source. The source was elevated most of the time, but it was lowered for shielding when necessary for research activities in the field such as planting, making observations and collection of materials for study in the lab. This method, termed the Gamma Field, was first used for a small source of approximately 16 curies in 1949 and 1950 (Sparrow and Singleton 1953), but over the years larger sources were installed. During the mid-1960s the source was increased to 3,160 curies in a field of about 5.2 hectares (12.8 acres) (Sparrow 1966). The purpose of much of this work was to intentionally generate mutations in order to understand the process of their formation. During this time period some researchers hoped positive mutations that would improve crop species could be created through experimental irradiation. This was not the main purpose of Sparrow's research at the BNL, but it was of interest as part of agricultural research funded by the AEC (Curry 2016).

Sparrow became a leader in the field of radiobiology. He was elected president of the Radiation Society, he started the journal *Radiation Botany* in1961 (renamed *Environmental and Experimental Botany* in 1976) and he published more than 300 scientific papers during his career. He is perhaps best known for establishing a relationship between radiosensitivity of a species and the size of the nucleus within the cells of the species (Sparrow and Evans 1961). The relationship emerged when many species were surveyed, especially in the research undertaken in the outdoor Gamma Field. Log linear regressions were found between measures of radiosensitivity versus nuclear size when the data was plotted. An example of this plot is given in Figure 8.1 for nuclear volume and dose of radiation causing growth inhibition.

Many other parameters for radiation impact could be plotted and, for example, another version of this type of relationship was given by Sparrow et al. (1963) for the lethal dose versus nuclear volume. Other factors were found to modify this relationship but it was an important guide for research and it was examined and tested at a number of study sites in the AEC's Radioecology Program.

George Woodwell Is Hired at the BNL

As the cobalt radiation source was increased over the years for research on cultivated plants, damage to the pine trees on the edge of the Gamma Field became apparent. This observation led Sparrow and his associates to consider the radiosensitivity of native trees and, ultimately, to wonder about the potential ecological effects of ionizing radiation. At this time in the late 1950s there was growing

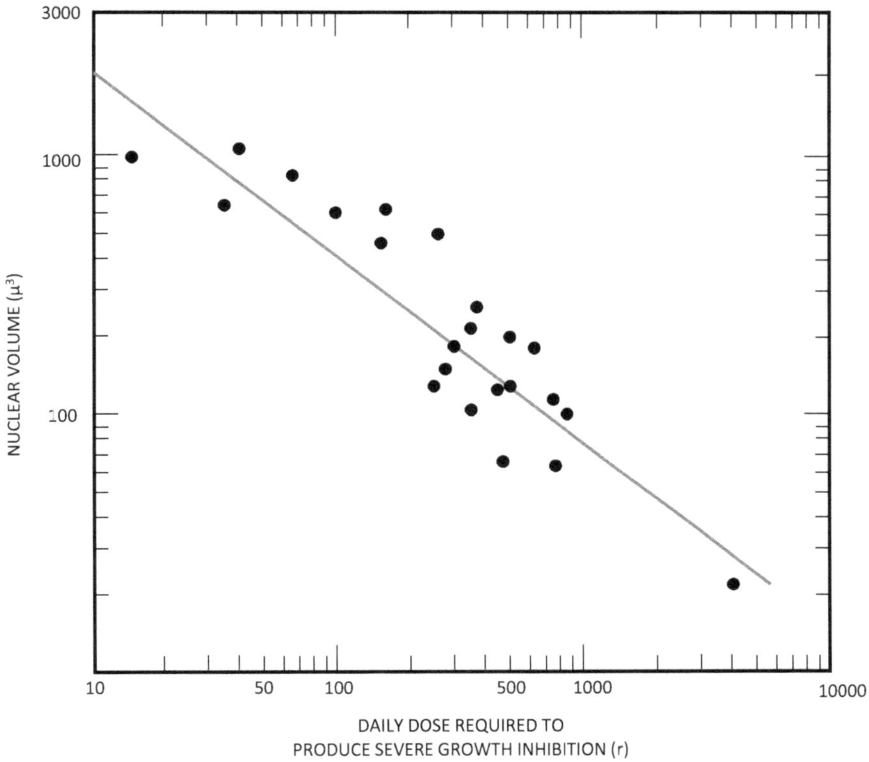

FIGURE 8.1 Relationship between nuclear volume and radiosensitivity in terms of growth inhibition (redrawn from Sparrow and Miksche 1961). The regression equation was based on analysis of 23 higher plant species.

interest in ecological studies within the AEC and Sparrow was able to secure funding for a new project to irradiate the oak-pine forest at the BNL.

George Woodwell was hired to head the Radioecology Project in what was a very fortuitous choice. Woodwell was creative and energetic and he had (and still has) big ideas about ecology. He was well qualified to join Sparrow's biology group at the BNL and to extend the research on radiosensitivity to the ecosystem scale. His doctoral work had been at Duke University under the direction of H. J. Oosting on the ecology of a pine species (Woodwell 1958). After leaving Duke he taught and conducted research at the University of Maine from 1957 to 1961. There he studied indirect effects of the insecticide DDT, which was routinely sprayed on the boreal forest to control spruce budworm outbreaks. This research covered the persistence of DDT in the forest soils and the effects of the toxin on reproduction and survival of the coniferous spruce and fir tree species (Woodwell 1962a, Woodwell and Martin 1964). DDT was shown to be a stressor on the forest, and this work provided good preparation for Woodwell's future work on the stress of ionizing radiation on the BNL forest. He would later continue research on the ecological effects of DDT and generalize his knowledge and experience to different kinds of stressors throughout his career.

Woodwell joined the Biology Department at the BNL in 1961 and would work there until 1975. He immediately became involved in Sparrow's ongoing research on the effects of radiation on the pitch pine (*Pinus rigida*) surrounding the Gamma Field (Sparrow et al. 1965, Woodwell and Miller 1963), while he helped to plan the forest irradiation experiment.

The Irradiation Experiments

In 1961 no one had yet set up and carried out a major comprehensive terrestrial ecosystem study essentially anywhere. Efforts were underway at several other AEC study sites but Woodwell took advantage of his new job to start such a study in the oak-pine forest at Brookhaven. He helped plan the forest irradiation experiment that Sparrow had proposed but he also expanded the scope of the work to include irradiation of an early successional old field ecosystem. Woodwell envisioned the research as an opportunity to extend Sparrow's approach of studying radiosensitivity of plant species to the ecosystem scale. Thus, by conducting simultaneous irradiation experiments in both an early successional system (the old field) and a late successional system (the forest), it would be possible to make comparative statements about radiosensitivity across a time series of ecosystems. The concept of ecological succession was fresh in Woodwell's mind when he arrived at Brookhaven. He had learned the importance of succession, which he later called (Woodwell 1967a) "one of the great central principles of biology", during his doctoral studies with H. J. Oosting, who had conducted important plant succession research in the mid-Atlantic region of the U.S. (Billings 1970, Oosting 1942). In terms of plant communities in the region, the stages of succession were known to be very different: early successional systems were dominated by fast-growing herbaceous species characterized by relatively low biomass, while later successional systems—the terminal stage being termed "climax"—were dominated by an overstory of trees and an understory of shrubs and herbs characterized by relatively high biomass. Based on these fundamental differences in ecosystem structure, Woodwell suspected that the ends of the successional time series (old field and forest) would differ in radiosensitivity in interesting ways.

Both irradiation experiments were set up in the same fashion, being modeled after the way Sparrow had set up the Gamma Irradiation Field facility. The radiation source was placed in a centrally located tower and it was able to be raised for irradiation or lowered for shielding by operation of a winch in a remote building. Routinely the source was raised for continuous exposure for 20 hours daily and it was lowered and shielded for four hours daily so that on-site research activities could be carried out. The sites were fenced to restrict access and appropriate safety procedures were instituted along with a dosimetry regime. The main research approach was to take measurements and plot parameters along the radiation gradient that extended from near the source with the highest exposure outward to a point where the radiation levels matched background levels. For the forest, the end of the exposure gradient at background levels was considered to be the control for the experiment. The old field site was located on the edge of the Gamma Irradiation Field and it was irradiated with a 3100 curie Co-60 source in 1962 and a 3900 curie Co-60 source in 1963. The forest was irradiated with a cesium source of 9500 curies that was installed in November 1961. The irradiated old field was studied intensively for only two years, but the results from the study provided important context for the overall work. The forest was irradiated for 15 years, which made it the longest-running experiment in the history of Radioecology. Details of the experimental design of the irradiated forest are given by Woodwell (1963) and Woodwell and Hammond (1962).

The Old Field

The old field was plowed at the end of the growing season in September 1961 and then allowed to revegetate naturally from locally available seed sources. Detailed studies were made of the vegetation that developed on the irradiated site and at a control site (Woodwell and Oosting 1965). Interestingly, Woodwell recruited the son of his Ph.D. advisor, Jan Oosting, to help with the research. A pattern of zonation developed outward from the radiation source as follows:

> The wide range of radiation exposures available for this experiment induced correspondingly wide divergence in the structure of old field communities. The highest exposures excluded all higher plants, while exposures of several hundred to a few thousand roentgens per day allowed development of simplified communities containing one or a few species. At still lower exposures more complex communities developed frequently as rich in species as the non-irradiated controls, but of different composition.
>
> *(Woodwell and Oosting 1965)*

The diversity of plant species decreased along the radiation gradient from the outward edge of the experimental plot to the source in the middle, as expected. However, biomass of the system behaved differently (Figure 8.2). No higher plants grew near the source, as expected, but a peak in biomass occurred relatively near the source, which was dominated by *Digitaria* spp. (crabgrass). This species was highly radioresistant in part because of its prostrate life-form, which caused it to grow close to the ground with less exposure to radiation than experienced by species with upright life-forms. *Digitaria* also had a relatively low chromosome volume, which further reduced its sensitivity to radiation (see Table 5 in Woodwell and Oosting 1965). In addition to these qualities that conferred radioresistance to *Digitaria*, the species may have benefited from increased access to resources that would have been shared with other competing species if there had been no stress.

Anomalies in growth and development of several plant species were noted along the radiation gradient. These are described in a section of the paper headed

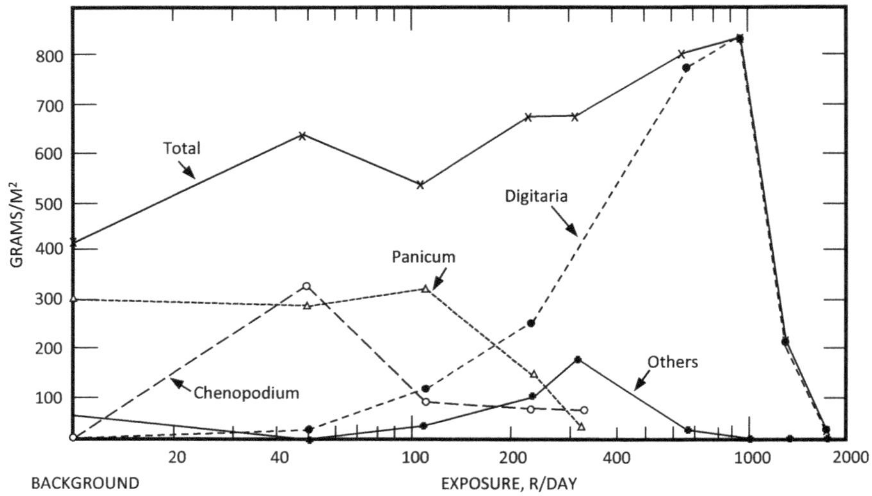

FIGURE 8.2 Distribution of aboveground biomass in the irradiated old field along the exposure gradient, showing the relative contribution of plant species to the total (from Woodwell 1967a).

Source: Used with permission of the AAAS.

"Teratology". For two species, *Chenopodium album* and *Amaranthus retroflexus*, height growth was reduced and the density of flowering heads was less at exposures greater than 50 r/day; and at exposures from 100 to about 300 r/day, the individuals were "highly fasciated and bushy in appearance, frequently having a flattened stem cross section, an abundance of flowers and unusually long leaves." Malformations were also found in *Erigeron canadensis*, which had reduced height growth, increased branching and asymmetric growth of stems in cross section. These plants actually had more seed production but most were not viable. Photos of growth anomalies caused by exposure to radiation are given by Woodwell and Oosting (1965).

The Forest

The main emphasis of radioecology at the BNL was the chronic irradiation of the oak-pine forest ecosystem. This study lasted 15 years and resulted in a number of publications describing radiation effects on different parts of the ecosystem. Although he was working at a large, well-funded research laboratory, Woodwell had minimal staff support to work on the research. Instead he drew on a network of colleagues and graduate students from nearby universities. A positive synergy developed between Woodwell and Sparrow during the early stages of the project that set it on a successful path. Woodwell and Sparrow were at different stages of their careers in 1961 when the project began. Woodwell arrived at the BNL as a relatively young scientist with little professional experience, while Sparrow was an established scientist with a national reputation in the field of radiobiology.

Woodwell was 33 years old when he arrived; Sparrow was 14 years older. Despite their differences, the two scientists collaborated well and published several papers together as coauthors. For his part, Woodwell had great respect for Sparrow, as evidenced when he stated, "Sparrow's work was extraordinary, exciting, and fundamental" in his memoir (Woodwell 2016).

Perhaps their most significant collaboration was represented by a pair of papers published at the beginning of the project that dealt with predictions of the effects of irradiation on the plant community of the forest (Sparrow and Woodwell 1962, Woodwell and Sparrow 1963). The first paper was based on a presentation by Sparrow at the first National Symposium of Radioecology. The paper was published both in the proceedings of the symposium (Sparrow and Woodwell 1963) and in the journal that Sparrow edited (Sparrow and Woodwell 1962). The majority of that paper covered the research that Sparrow had led on radiosensititivity based on characteristics of a species' nucleus. At the end of the paper, predictions are made on how dominant plant species in the forest would respond to irradiation. These predictions were presented in a diagram based on the circular research plot showing the health status of species along the radiation gradient after one year of exposure. The predictions were made based on measures of nuclear volumes along with assumed growth-response curves from Sparrow's experience with the Gamma Irradiation Field. The goal of the effort was to predict "the gross physiognomy to be expected as a result of one year's exposure of the oak-pine forest to chronic gamma irradiation and to make certain generalizations about effects of longer exposure" (Sparrow and Woodwell 1962). The second paper (Woodwell and Sparrow 1963) was shorter than the first (Sparrow and Woodwell 1962) and focused solely on results of the irradiation experiment after nine months for six of the dominant woody species—three trees and three shrubs. Graphs are included, showing the projected and observed growth of the plant species along the radiation exposure gradient. In all six cases the growth of the plants was less than had been predicted, demonstrating that they were more sensitive to irradiation than had been expected. In the discussion of the paper, it was suggested that indirect effects associated with competitive interactions between the species may have modified the direct effects of irradiation, which had been the basis of the original predictions using Sparrow's nuclear volume theory of radiosensitivity. Taken together, these two papers are important in hinting at the complexity of the ecosystem. The first paper is significant because the authors were able to essentially make predictions about the ecosystem scale based on the dimension of the nucleus at the cellular scale. The second paper is significant because it demonstrates that information from the cellular scale is not sufficient to predict ecosystem responses and that more knowledge of ecological interactions is needed to understand the effects of irradiation of whole ecosystems in the field.

Overall, the irradiation of the forest quickly and dramatically affected the ecosystem as anticipated. Within six months after installation of the cesium source, mortality from radiation stress created a series of obvious zones of vegetation around the source that persisted throughout the 15-year experiment. As described by Gochenaur and Woodwell (1974), the zones "consisted of a central devastated zone where no higher plants survived although certain radioresistant lichens and mosses persisted, a Sedge Zone, a Shrub Zone, an Oak Forest Zone and the intact Oak-Pint forest at distances greater than 100 m". The zonation pattern was the result of a nonrandom sorting of species in the forest caused by the direct effect of irradiation along with indirect effects due to associated ecological interactions. The pattern was "closely related to the stratification of the vegetation, with the tree stratum the most sensitive and the ground cover layer the most resistant" (Woodwell and Rebuck 1967). Lichens were the organisms most resistant to irradiation, with 11 species found near the source with an exposure of 2250 R/day at a density of two thalli/m2 (Brodo 1964, Woodwell and Gannutz 1967). Thus, radiosensitivity was found to be broadly correlated with the stature or height of the vegetation. According to Woodwell (1967a), irradiation had caused "a systematic dissection of the forest, strata being removed layer by layer." This result was similar to that found when the old field was irradiated: the upright species were more sensitive than the low-growing species. Woodwell (1967a) later related the Brookhaven zonation pattern to other well-known examples of zonation:

> The sorting by size, which now seems to be a well-established characteristic of radiation damage, has interesting parallels elsewhere in nature. It occurs along gradients of increasing climatic severity, such as the transition from forest to tundra in the north, and on mountain slopes.

Woodwell published a series of summary papers as the experiment continued over the years (Woodwell 1962b, 1963, 1967a, Woodwell and Whittaker 1968a, Woodwell and Houghton 1990); these papers represent informal progress reports and they record the evolution of his thinking as it developed over time. This series of papers also demonstrates the priority that Woodwell gave to communicating the results of irradiation experiment to a wide audience. Many diagrams and photographs included in the series of summary papers illustrate what he came to describe as the "impoverishment" of the forest by chronic gamma irradiation (Woodwell and Houghton 1990).

Unlike the old field, when the forest was irradiated, both diversity and biomass decreased along the exposure gradient from the outer edge of the study plot inward to the cesium source (Figure 8.3).

While many species were killed by radiation stress, there were several interesting examples of successful species that emerged from the pool of existing species in the forest along the radiation gradient, similar to the success of *Digitaria* in the old field irradiation experiment already mentioned. One of the most successful species that emerged along the radiation gradient was the sedge *Carex pennsylvanica*, which formed a nearly monoculture zone 20 to 40 meters (65 to 130 feet) from the radiation source. This species is normally an inconspicuous part of the forest understory that has been described as "an ubiquitous but sparse herb" (Woodwell 1967a). After irradiation of the forest began, *Carex* expanded vigorously by vegetative growth. As described by Wagner (1966), "The species is uniquely suited

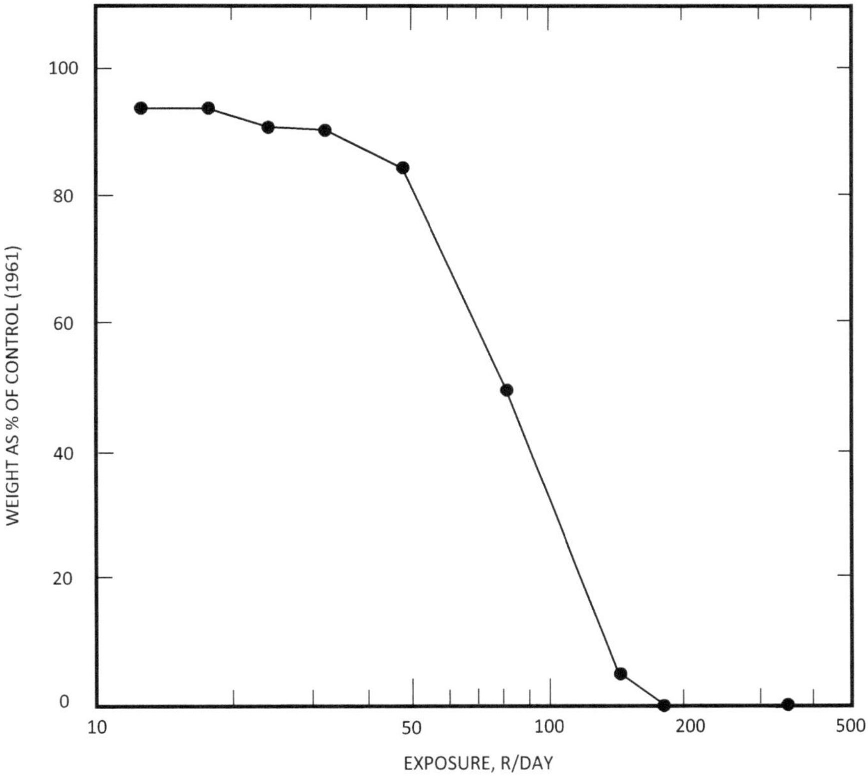

FIGURE 8.3 Distribution of total aboveground biomass in the irradiated forest along the exposure gradient (from Woodwell 1967a).

Source: Used with permission of the AAAS.

to a radiation environment by a shielded perennating bud, vigorous production of rhizomes stimulated by cropping, ability to penetrate litter where present and finally a low interphase chromosome volume indicative of innate resistance to radiation." This species has a low, graminoid growth form and it grew to form a dense and almost continuous mat along the radiation gradient. Its success was probably due both to its relative radioresistance and because mortality of the woody tree and shrubs increased available sunlight and reduced root competition. Thus, within the irradiated forest *Carex* emerged from the existing pool of understory species and, as described by Wagner (1966), "filled the radiation induced vegetational vacuum" for a zone 20 to 40 meters (65 to 130 feet) from the radiation source.

Beyond the *Carex* monoculture at lower exposure rates, a zone of low (0.3–1.0 meter or 1.0–3.2 feet tall) ericaceous shrubs emerged in the irradiated forest. This zone was dominated by black huckleberry (*Gaylussacia baccata*) and low blueberry (*Vaccinium vacillans*), which formed a zone 40–50 meters (130–160 feet) from the radiation source. Both of these species regenerate by sprouting from their underground rhizome systems, which allowed them to spread vegetatively along the radiation gradient.

Vaccinium was found to be more radioresistant than *Gaylussacia*, even though the volume of its nuclei were larger and therefore more susceptible to radiation damage, because its rhizome system grew deeper in the soil and thus was better protected from radiation exposure. These shrub species likely benefited from the mortality of the tree overstory, as mentioned for *Carex*, by having more sunlight and reduced root competition. Interestingly, Brayton and Woodwell (1966) suggested the success of these shrub species under irradiation was at least due to adaptation that evolved in response to fire, which is a natural, recurring disturbance to the mid-Atlantic oak-pine forest.

Woodwell and Houghton (1990) introduced the term "sorting" to describe the self-organizing processes that generated the zonation pattern around the radiation source and the relative success or failure of species in response to radiation stress. They summarized the sorting of vegetation along the radiation exposure gradient as a two-stage process:

> [T]he first sorting was the winnowing of the species of the original forest by chronic exposure to ionizing radiation. . . . The effects were the combination of direct effects on species and the effects of the change in the structure of the forest brought by these direct effects. The combination produced the patterns outlined.

The second sorting involved "the slow accumulation of hardy invaders that became established in each zone." He termed the invaders as "adventives" or "hardy plants, commonly found in heavily disturbed impoverished sites and also resistant to ionizing radiation. The reason for the radiation resistance is surely one of the most remarkable facts of nature" (Woodwell 1983).

A number of interesting examples of species' success (populations increased) or failure (populations decreased) are described in the detailed inventories of the herbaceous understory in the irradiated forest done by Wagner (1965, 1966) from 1963 to 1965 and by Flaccus et al. (1974) in the years 1968–1973.

Many aspects of the structure and function of the irradiated forest were examined throughout the period of chronic exposure to the cesium source. Some of the early studies established baseline conditions for the study and are important contributions to the quantification of ecosystem performance. The commitment to studying chronic irradiation over more than a decade allowed long-term responses of the forest to be differentiated from short-term responses as the research continued. Interestingly, even after 15 years of study, no equilibrium was reached in terms of species turnover (Woodwell and Houghton 1990).

Perhaps of most significance in the research approach was the reliance on studying populations and ecosystem parameters in terms of changes along the gradient in radiation exposure from the source outward to the point where only background radiation occurred. These studies were complicated by changing conditions inherent in a large-scale radiation experiment in nature. For example, Cesium-137 has a half-life of 30 years, so the source became less radioactive over time. This change in intensity of radiation exposure required periodic recalibration of the dosimetry of the study plot due to the long-term nature of the research. Also, shielding altered the responses along the radiation gradient, as described by Woodwell and Rebuck (1967):

> The principal variability other than distance was introduced by shielding from the stems of trees, which reduced exposures by as much as 50%. A rule of thumb useful in such work is that there is a 5% reduction in dose for each centimeter of tissue intervening in a direct line to the source. The net effect of this shielding was to allow survival of shielded plants or portions of plants at distances where the average exposures were in fact lethal for those species.

Shielding effects could be dramatic; for example Woodwell and Whittaker (1968a) referred to the *Carex* plants living behind tree trunks as "green shadows" in the zone where most higher plants had been killed by irradiation. Despite these complications, the gradient approach of reporting data was a powerful way of depicting the effects of radiation in the forest. Examples of the many studies are listed in Table 8.1, which demonstrates the breath of research undertaken in the Brookhaven irradiated forest.

Although the research program was broad, not much attention was given to animals in the forest. For example, only one of the 11 citations given in Table 8.1 relates to

TABLE 8.1 Listing of studies done in the Brookhaven irradiated forest over time with notation of a representative radiation gradient graph.

Date of Study	*Example of Gradient*	*Reference*
1962–1966	Figure 1. Total litterfall	Woodwell and Marples (1968)
1963–1965	Figure 5. Plant species diversity	Woodwell and Rebuck (1967)
1963	Figure 2. Soil moisture as an index of ET	Reiners and Woodwell (1966)
1963	Figure 1. Aphid population outbreak	Woodwell and Brower (1967)
1964	Figure 1. Ground cover with fire comparison	Brayton and Woodwell (1966)
1964	Figure 3. Diversity of lichen growth forms	Woodwell and Gannutz (1967)
1967	Figure 1. Prokaryotic vs. eukaryotic soil algae	Franz and Woodwell (1973)
1968	Figure 1. Soil organic matter content	Horrill and Woodwell (1973)
1968	Figure 1. Microfungal abundance	Gochenaur and Woodwell (1974)
1969	Figure 1. Macrofungi growth performance	Holt (1972)
1973 (also 1968)	Figure 2. *Carex* distribution over time	Flaccus et al. (1974)
1973	Figure 1. Total litterfall	Armentano and Woodwell (1976)

animals. However, the study for the aphid outbreak (Woodwell and Brower 1967) is significant. In the summer of 1963, about a year and a half after irradiation began, the population of the aphid *Myzocallis dicolor* erupted to several hundred times its normal density. The outbreak occurred on oak trees and ranged from about 1,000 aphids/m2 at 5 R/day to 350,000 aphids/m2 at 9 R/day. Thus it occurred over a narrow portion of the radiation gradient in the irradiated forest and it was not observed in the surrounding undamaged forest. The cause of the outbreak was unknown, but it seemed to be triggered by irradiation, perhaps due to changes in the chemistry of the damaged oak leaves or by decreased pressures from predators, parasites or competitors. This study was part of the Ph.D. dissertation research of Brower (1965), who also reported on changes in an ant colony within the forest that received high levels of radiation exposure (Brower 1966). Additionally, a study of small mammals was conducted in 1961 before and after irradiation began (Bongiorno and Pearson 1964). Seven species were trapped with the white-footed mouse, *Peromyscus leucopus*, as the dominant. No effect of radiation on the distribution of the small mammals could be detected, at least in part due to problems with sampling methods. Beyond these studies, anecdotal references were made on rabbit herbivory on herbaceous plants along the radiation gradient. It was suggested that rabbits were attracted to the clearing in the forest caused by irradiation and that their grazing pressure in the *Carex* zone and the shrub zone affected the herbaceous plants (Wagner 1966, Flaccus et al. 1974).

To conclude the review of radiation impacts in the forest, few mutant organisms were found during the detailed inventories of the biodiversity of the ecosystem. The only reported examples are nine taxa of soil microfungi out of 267 taxa that were found by Gochenaur and Woodwell (1974) during fieldwork in 1968. These mutant taxa were found with aberrant sporulating structures or changes in colony color (typically loss of pigmentation). This relative lack of mutants in the irradiated forest is a bit ironic given that Sparrow's research group in the nearby Gamma Field was intentionally trying to create mutants that might have improved agronomic qualities! However, the paucity of surviving mutant organisms was commonly found in studies of radiation effects on ecosystems. Woodwell (1970a) summarized his views on mutagenic pollutants, such as radionuclides, as follows:

> [Radionuclides] present a specific, direct hazard to man. They are much more important to man than to other organisms. A slightly enhanced rate of mutation is a serious danger to man, who has developed through medical science elaborate ways of preserving a high fraction of the genetic defects in the population; it is trivial to the rest of the biota, in which genetic defects may be eliminated through selection. This is an important fact about pollution hazards—toxic substances that are principally mutagenic are usually of far greater direct hazard to man than to the rest of the earth's biota and must be considered first from the standpoint of their movement to man through food webs or other mechanisms and to a much lesser extent from that of their effects on the ecosystem through which they move.

General Ecology at the BNL

In the mid-1960s Woodwell developed a track of research parallel to Radioecology with studies of forest productivity and metabolism. This new emphasis grew out of work being done in the irradiated forest, but it dealt solely with analysis of the undisturbed forest, without the influence of the Cesium source. This work began early in the project through collaboration with P. F. Bourdeau of Yale University. Together they published papers on carbon dioxide exchange of pine branches in small closed chambers (Bourdeau and Woodwell 1964, 1965) and biomass production in the irradiated forest (Woodwell and Bourdeau 1965). Woodwell's interest is also illustrated by the organic matter budget given in Woodwell and Marples (1968), which is an early attempt to quantify the inputs and outputs at the system scale (Figure 8.4). This early ecosystem diagram focuses on litter layer processes but it was later enlarged to encompass the whole system as more data were synthesized from the project. Specifically in terms of Radioecology, the CO_2 exchange data was not found to provide an adequate indicator of radiation damage, but the work stimulated Woodwell to continue to develop methods for estimating productivity and metabolism as indices of forest system performance (see Woodwell and Dykeman 1966 as an example).

An important catalyst to this line of research was the collaboration starting in 1964 with Robert Whittaker, who was a professor at Brooklyn College, in New York City. Whittaker had been developing techniques for estimating net production of eastern deciduous forests (see, for example, Whittaker 1961, 1963, 1966) that were similar to those being formulated by Woodwell, so the

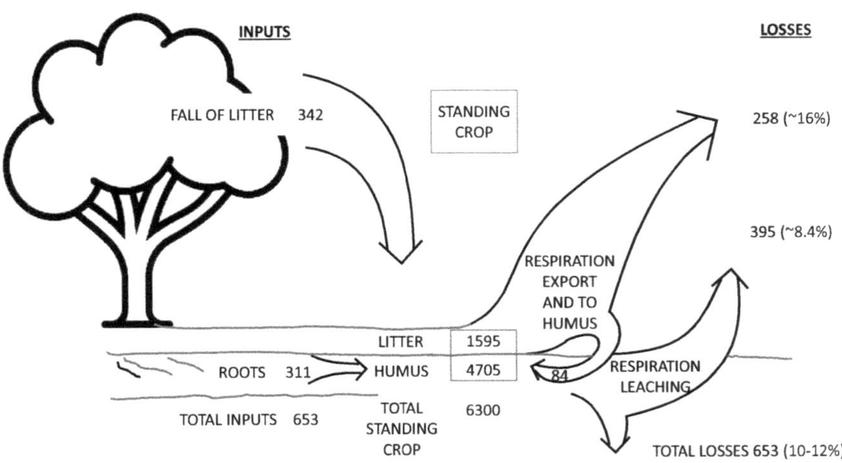

FIGURE 8.4 Cross-sectional diagram of organic matter flows in the forest soil at Brookhaven, New York (from Woodwell and Marples 1968). Data are in grams dry weight per square meter per year.

Source: Used with permission of the Ecological Society of America.

connection was natural. Westman and Peet (1985) in their biographical sketch of Whittaker provide the following summary of the collaboration:

> In 1964 George M. Woodwell persuaded Whittaker to take a year's leave to work with him at Brookhaven National Laboratory. The partnership was electric, in part because the two men shared a commitment to high standards and hard work, and perhaps also because Woodwell's humorful and more expansive style well complemented Whittaker's more intense one. The two developed a profound respect and fondness for each other, and collaborated successfully on eight papers on the Brookhaven oak-pine forest and surrounding vegetation.

Together, Woodwell and Whittaker developed what became a standard technique termed dimension analysis (Woodwell and Whittaker 1968b), which utilized allometric relationships to estimate forest biomass and production from basal diameters of trees and other easily measured features. They applied this technique, which Whittaker and Niering (1975) later called the "Brookhaven System", to the oak-pine forest at the BNL and found net primary production to be 1195 grams dry weight/m2/year (Whittaker and Woodwell 1968, 1969). Woodwell also had continued to develop a series of gas exchange techniques, in what was termed the "Brookhaven Approach" (Woodwell and Botkin 1970), which provided an independent measurement of the forest respiration rate. This estimate was combined with the dimension analysis results to calculate the gross primary production of the forest at about 2600 grams dry weight/m2/year (Whittaker and Woodwell 1969). The culmination of this line of nonradioecology research was the chapter in the famous "Biosphere" issue of the magazine *Scientific American* that included a detailed model of organic energy flows in the Brookhaven oak-pine forest (Woodwell 1970b). Of added significance, this chapter also included an excellent plate of photographs showing images of the radiation zones within the experimental forest.

Heritage

Woodwell left the BNL and Radioecology research in 1975 when he founded and became director of the Ecosystem Center at the Marine Biological Laboratory in Woods Hole, Massachusetts. This center was supported by foundations and grants, and it focused on research on local and global environmental problems. He stayed there for 10 years, then left and founded the Woods Hole Research Center, with a mission similar to the Ecosystem Center, where he was director from 1985 until his retirement in 2005. From 1977 to 1978 he served as president of the Ecological Society of America (Anonymous 1977). It is no coincidence that the chronic irradiation experiment in the oak-pine forest at the BNL ended in 1976, after Woodwell's departure. The forest had been irradiated by the cesium source for 15 years.

In some autobiographical writing, Woodwell (2009, 2016) candidly and respectfully described his time at the BNL. He valued the opportunity to supervise the Radioecology project, but it is also clear that his interests in environmental issues were broader than this single problem. Even before he left the BNL, Woodwell had begun research on other topics. For example, he led an important project on biomagnification of insecticides in a Long Island salt marsh (Woodwell et al. 1967). He also organized and contributed to symposia that became benchmarks in their fields: one entitled Diversity and Stability in Ecological Systems, in 1969 (Woodwell and Smith 1969), and another entitled Carbon and the Biosphere, in 1973 (Woodwell and Pecan 1973). Always an effective writer and communicator, Woodwell wrote a number of general papers on pollution (Woodwell 1967b, 1970a, 1981, Woodwell et al.1966). In regard to Radioecology, his paper on fallout in the journal *BioScience* (Woodwell 1969) was reprinted multiple times in subsequent environmental compendia (see, for example, Anderson 1971, Dansereau 1970). Woodwell's early experiences with measuring CO_2 concentrations and estimating the organic balance of the oak-pine forest at the BNL were an appropriate background for his transition into research on the global carbon balance after his departure (see, for example, Woodwell et al. 1978, Woodwell 1995).

It is unclear if Woodwell and the 15-year Radioecology project had any lasting influence on the BNL. The lab seems not to have hired another ecologist after Woodwell left. Furthermore, the experimental forest did not become a long-term research site, as did occur at the conclusion of some other AEC-supported Radioecology projects, such as at El Verde, Puerto Rico, or Oak Ridge, Tennessee. However, several follow-up studies of the forest have been conducted with interesting results (Olsvig 1979, Stalter and Kincaid 2009). In terms of data, the BNL forest productivity analysis led by Woodwell has been cited in comparative summaries many times (see, for example, Reichle 1981, Aber and Melillo 2001), demonstrating another component of the heritage of the radioecology project.

A very significant part of the heritage of the radioecology project was the large number of students who did their graduate research at the BNL. These students came from Rutgers, in New Jersey; the University of Massachusetts at Amherst; Clark University, in Massachusetts; the University of Illinois; and Duke University, in North Carolina. It is a bit remarkable that Woodwell was able to recruit students from such a broad area, but it reflects the network of collaborators he developed and, of course, his charisma and leadership skills. Many of the students who worked on the BNL Radioecology Project went on to distinguished and productive careers in environmental science. A few examples are William Reiners, who became a leader in the field of Ecosystem Ecology and who received the Eminent Ecologist Award from the Ecological Society of America in 2013, and Daniel Botkin, who developed the first ecosystem-scale computer model of a forest (Botkin et al. 1972, Botkin 1993), and who has authored or coauthored at least 12 books on aspects of environmental science and natural history. As an aside, Botkin included a chapter in one of his books (Botkin 2003) on his personal experiences as a graduate student working on the BNL Radioecology Project entitled "The Radioactive Forest".

A final contribution to heritage of the Brookhaven Radioecology Project relates to assessment of the potential ecological impact of nuclear war on ecosystems. Many of the scientists who conducted AEC-supported, large-scale ecosystem experiments in Radioecology in the 1960s and early 1970s were reluctant to discuss nuclear war implications of their research, even though that was one of, if not the, dominant public issues of the Cold War era. Woodwell, on the other hand, seemed to have felt socially and/or morally compelled to discuss the subject throughout his career. He acknowledged that the chronic irradiation of the BNL forest did not simulate the effects of nuclear war or even a reactor accident, which were better modeled by experiments of acute, or short-term, exposure, as occurred at El Verde, Puerto Rico. But he strived multiple times to relate the results of the Gamma Irradiated Forest Experiment to nuclear war (see Woodwell and Sparrow 1965, the last paragraph in Woodwell and Rebuck 1967, Woodwell and Holt 1971). Of significance, in 1963 he organized a symposium entitled Ecological Effects of Nuclear War, and he edited the proceedings of the conference (Woodwell 1965). This was a very early effort to imagine the ecological consequences of a nuclear war, and was ahead of its time as an environmental impact assessment. Woodwell and Sparrow (1965) contributed an important paper to the symposium that projected three categories of damage to natural ecosystems (minor effects, intermediate effects and severe effects) oriented around the concept of succession that would occur after a nuclear attack. They suggested "probably the most important single factor that might delay recovery indefinitely in North America would be unstable soils" due to erosion and radiation damage. The symposium included five papers along with a summary by E. P. Odum. The paper by Robert Platt in the symposium, entitled "Ionizing Radiation and Homeostasis of Ecosystems", was particularly interesting in summarizing the early Radioecology that had been done at Emory University. In general, all of Woodwell's early writings on nuclear war provided background for the major theme of nuclear winter, which became an important subject of the 1980s among ecologists and other scientists (Woodwell 1983, 1985).

Literature Cited

Aber, J. D. and J. M. Melillo. 2001. Terrestrial Ecosystems, Second Edition. Academic Press, San Diego, CA.

Anderson, P. K. 1971. Omega, Murder of the Ecosystem and Suicide of Man. W. C. Brown Co., Dubuque, IA.

Anonymous. 1977. George M. Woodwell, President 1977–1978. Bulletin of the Ecological Society of America 58(3)7.

Armentano, T. V. and G. M. Woodwell. 1976. The production and standing crop of litter and humus in a forest exposed to chronic gamma irradiation for twelve years. Ecology 57:360–366.

Billings, D. 1970. Henry J. Oosting 1903–1968. Bulletin of the Ecological Society of America 51(4):16–17.

Bongiorno, S. F. and P. G. Pearson. 1964. Orientation of *Peromyscus* in relation to chronic gamma radiation and vegetation. American Midland Naturalist 72:82–92.

Botkin, D. B. 1993. Forest Dynamics: An Ecological Model. Oxford University Press, Oxford, UK.

Botkin, D. B. 2003. Strange Encounters, Adventures of a Renegade Naturalist. Penguin Publ., New York, NY.

Botkin, D. B., J. F. Janak and J. R. Wallis. 1972. Some ecological consequences of a computer model of forest growth. Journal of Ecology 60:849–872.

Bourdeau, P. F. and G. M. Woodwell. 1964. Field measurements of carbon dioxide exchange by *Pinus rigida* trees exposed to chronic gamma irradiation. Ecology 45:403–406.

Bourdeau, P. F. and G. M. Woodwell. 1965. Measurements of plant carbon dioxide exchange by infra-red absorption under controlled conditions and in the field. Pp. 283–289. In: F. E. Eckardt (ed.). Methodology of Plant Eco-Physiology. UNESCO, Paris.

Brayton, R. D. and G. M. Woodwell. 1966. Effects of ionizing radiation and fire on *Gaylussacia baccata* and *Vaccinium vacillans*. American Journal of Botany 53:816–820.

Brodo, I. M. 1964. Field studies of the effects of ionizing radiation on lichens. The Bryologist 67:76–87.

Brower, J. H. 1964. Changes in the insect populations of the low shrub synusia in a natural forest community exposed to chronic gamma radiation stress. M. S. Thesis, University of Massachusetts, Amherst, MA.

Brower, J. H. 1965. Ecological changes in the insect community of a pitch pine-oak forest subjected to chronic gamma irradiation. PhD Dissertation, University of Massachusetts, Amherst, MA.

Brower, J. H. 1966. Behavioral changes in an ant colony exposed to chronic gamma irradiation. American Midland Naturalist 75:530–534.

Cohn, S. H., J. S. Robertson and R. A. Conrad. 1960. Radioisotopes and environmental circumstances: The internal radioactive contamination of a Pacific Island community exposed to local fallout. Pp. 306–330. In: R. S. Caldecott and L. A. Snyder (eds.). A Symposium on Radioisotopes in the Biosphere. University of Minnesota, Minneapolis, MN.

Conger, A. D. 1977. Arnold Hicks Sparrow (1914–1976). Radiation Research 69:194–196.

Curry, H. A. 2016. Atoms in agriculture: A study of scientific innovation between technological systems. Historical Studies in the Natural Sciences 46:119–153.

Dansereau, P. (ed.). 1970. Challenge for Survival. Columbia University Press, New York, NY.

Flaccus, E., T. V. Armentano and M. Archer. 1974. Effects of chronic gamma radiation on the composition of the herb community of an oak-pine forest. Radiation Botany 14:263–271.

Franz, E. H. and G. M. Woodwell. 1973. Effects of chronic gamma irradiation on the soil algal community of an oak-pine forest. Radiation Botany 13:323–329.

Gochenaur, S. E. and G. M. Woodwell. 1974. The soil microfungi of a chronically irradiated oak-pine forest. Ecology 55:1004–1016.

Holt, B. R. 1972. Radiosensitivity and distribution of common macro-fungi in a chronically irradiated oak-pine forest. Radiation Botany 12:339–342.

Horrill, A. D. and G. M. Woodwell. 1973. Structure and cation content of a podzolic soil of Long Island, N. Y., seven years after destruction of the vegetation by chronic gamma irradiation. Ecology 54:441–444.

Olsvig, L. S. 1979. Pattern and diversity analysis of the irradiated oak-pine forest, Brookhaven, New York. Vegetatio 40:65–78.

Oosting, H. J. 1942. An ecological analysis of the plant communities of Piedmont, North Carolina. American Midland Naturalist 28:1–126.

Reichle, D. E. (ed.). 1981. Dynamic Properties of Forest Ecosystems. Cambridge University Press, Cambridge, UK.

Reiners, W. A. and G. M. Woodwell. 1966. Water budget of a forest damaged by ionizing radiation. Ecology 47:303–306.
Sparrow, A. H. 1960. Uses of large sources of ionizing radiation in botanical research and some possible practical applications. Pp. 195–219. In: Large Radiation Sources in Industry, Volume 2. International Atomic Energy Agency, Vienna, Austria.
Sparrow, A. H. 1966. Research uses of the gamma field and related radiation facilities at Brookhaven National Laboratory. Radiation Botany 6:377–405.
Sparrow, A. H., R. L. Cuany, J. F. Miksche and L. A. Schairer. 1961. Some factors affecting the responses of plants to acute and chronic radiation exposures. Radiation Botany 1:10–34.
Sparrow, A. H. and H. J. Evans. 1961. Nuclear factors affecting radiosensitivity. I. The influence of nuclear size and structure, chromosome complement, and DNA content. Pp. 76–100. In: Fundamental Aspects of Radiosensitivity. Biology Department, Brookhaven National Laboratory, Upton, NY.
Sparrow, A. H. and J. P. Miksche. 1961. Correlation of nuclear volume and DNA content with higher plant tolerance to chronic radiation. Science 134:282–283.
Sparrow, A. H., L. A. Schairer and R. C. Sparrow. 1963. Relationship between nuclear volumes, chromosome numbers, and relative radiosensitivities. Science 141:163–166.
Sparrow, A. H., L. A. Schairer and G. M. Woodwell. 1965. Tolerance of *Pinus rigida* trees to a ten-year exposure to chronic gamma irradiation from Cobalt-60. Radiation Botany 5:7–22.
Sparrow, A. H. and W. R. Singleton. 1953. The use of radiocobalt as a source of gamma rays and some effects of chronic irradiation on growing plants. American Naturalist 87:29–48.
Sparrow, A. H. and G. M. Woodwell. 1962. Prediction of the sensitivity of plants to chronic gamma irradiation. Radiation Botany 2:9–26.
Sparrow, A. H. and G. M. Woodwell. 1963. Prediction of the sensitivity of plants to chronic gamma irradiation. Pp. 257–270. In: V. Schultz and A. W. Klement, Jr. (eds.). Radioecology, Proceedings of the First National Symposium on Radioecology. Reinhold Publishing Corp., New York, NY.
Stalter, R. and D. Kincaid. 2009. Community development following gamma radiation at a pine-oak forest, Brookhaven National Laboratory, Long Island, New York. American Journal of Botany 96:2206–2213.
Van't Hof, J. 1977. Obituary, Dr. Arnold Hicks Sparrow 1914–1976. Environmental and Experimental Botany 17:3.
Wagner, R. H. 1965. The annual seed rain of adventive herbs in a radiation damaged forest. Ecology 46:517–520.
Wagner, R. H. 1966. Herbaceous recolonization of the Brookhaven irradiated forest: The first three years. Radiation Botany 6:561–565.
Westman, W. R. and R. K. Peet. 1985. Robert H. Whittaker (1920–1980): The man and his work. Pp. 5–30. In: R. K. Peet (ed.). Plant Community Ecology: Papers in Honor of Robert H. Whittaker. Dr. W. Junk Publ., Dordrecht.
Whittaker, R. H. 1961. Estimation of net primary production of forest and shrub communities. Ecology 42:177–180.
Whittaker, R. H. 1963. Net production of heath balds and forest heaths in the Great Smoky Mountains. Ecology 44:176–182.
Whittaker, R. H. 1966. Forest dimensions and production in the Great Smoky Mountains. Ecology 47:103–121.
Whittaker, R. H. and W. A. Niering. 1975. Vegetation of the Santa Catalina Mountains, Arizona. V. biomass, production and diversity along the elevation gradient. Ecology 56:771–790.
Whittaker, R. H. and G. M. Woodwell. 1968. Dimension and production relations of trees and shrubs in the Brookhaven Forest, New York. Journal of Ecology 56:1–25.

Whittaker, R. H. and G. M. Woodwell. 1969. Structure, production and diversity of the oak-pine forest at Brookhaven, New York. Journal of Ecology 57:155–174.
Woodwell, G. M. 1958. Factors limiting the growth of pond pine seedlings in organic soils of the Carolinas. Ecological Monographs 28:219–236.
Woodwell, G. M. 1962a. Effect of DDT on cone production, germination, and seedling survival in the boreal forest. Ecology 43:396–403.
Woodwell, G. M. 1962b. Effects of ionizing radiation on terrestrial ecosystems. Science 138:572–577.
Woodwell, G. M. 1963. Design of the Brookhaven experiment on the effects of ionizing radiation on a terrestrial ecosystem. Radiation Botany 3:125–133.
Woodwell, G. M. (ed.). 1965. Ecological Effects of Nuclear War. BNL 917 (C-43). Brookhaven National Laboratory, Upton, NY.
Woodwell, G. M. 1967a. Radiation and the patterns of nature. Science 156:461–470.
Woodwell, G. M. 1967b. Toxic substances and ecological cycles. Scientific American 216(3):24–31.
Woodwell, G. M. 1969. Radioactivity and fallout: The model pollution. BioScience 19:884–887.
Woodwell, G. M. 1970a. Effects of pollution on the structure and physiology of ecosystems. Science 168:429–422.
Woodwell, G. M. 1970b. The energy cycle of the biosphere. Scientific American 223(3):64–74.
Woodwell, G. M. 1981. Toxic substances: Clear science, foggy politics. Pp. 5–20. In: B. W. Cornaby (ed.). Management of Toxic Substances in Our Ecosystems, Taming the Medusa. Ann Arbor Science Publ., Ann Arbor, MI.
Woodwell, G. M. 1983. The biotic effects of ionizing radiation. Pp. 131–139. In: J. Peterson (ed.). The Aftermath: The Human and Ecological Consequences of Nuclear War. Pantheon Books, New York, NY.
Woodwell, G. M. 1985. Energy enough to cleave the Earth. Bioscience 35:584–585.
Woodwell, G. M. 1995. Biotic feedbacks from the warming of the Earth. Pp. 3–21. In: G. M. Woodwell and F. T. Mackenzie (eds.). Biotic Feedback in the Global Climatic System, Will the Warming Feed the Warming? Oxford University Press, New York, NY.
Woodwell, G. M. 2009. The Nature of a House, Building a World that Works. Island Press, Washington, DC.
Woodwell, G. M. 2016. A World to Live In: An Ecologist's Vision for a Plundered Planet. MIT Press, Cambridge, MA.
Woodwell, G. M. and D. B. Botkin. 1970. Metabolism of terrestrial ecosystems by gas exchange techniques: The Brookhaven approach. Pp. 73–85. In: D. Reichle (ed.). Analysis of Temperate Forest Ecosystems. Springer-Verlag, NY.
Woodwell, G. M. and P. F. Bourdeau. 1965. Measurement of dry matter production of the plant cover. Pp. 519–527. In: F. E. Eckardt (ed.). Methodology of Plant Eco-Physiology. UNESCO, Paris.
Woodwell, G. M. and J. H. Brower. 1967. An aphid explosion induced by chronic gamma irradiation of a forest. Ecology 48:680–683.
Woodwell, G. M. and W. R. Dykeman. 1966. Respiration of a forest measured by carbon dioxide accumulation during temperature inversions. Science 154:1031–1034.
Woodwell, G. M. and T. P. Gannutz. 1967. Effects of chronic gamma radiation on lichen communities of a forest. American Journal of Botany 54:1210–1215.
Woodwell, G. M. and E. C. Hammond. 1962. A descriptive technique for the study of the effects of chronic ionizing radiation on a forest ecological system. BNL 715 (T-251). Brookhaven National Laboratory, Upton, NY.

Woodwell, G. M. and B. R. Holt. 1971. Effect of nuclear war on the structure and function of natural communities: An appraisal based on experiments with gamma radiation. Pp. 482–491. In: D. W. Bensen and A. H. Sparrow (eds.). Survival of Food Crops and Livestock in the Event of Nuclear War. US AEC, Office of Information Services, Washington, DC.

Woodwell, G. M. and R. A. Houghton. 1990. The experimental impoverishment of natural communities: Effects of ionizing radiation on plant communities, 1961–1976. Pp. 9–24. In: G. M. Woodwell (ed.). The Earth in Transition, Patterns and Processes of Biotic Impoverishment. Cambridge University Press, Cambridge, UK.

Woodwell, G. M., W. M. Malcolm and R. H. Whittaker. 1966. A-bombs, bugbombs, and us. Publication No. 9842, Brookhaven National Laboratory, Upton, NY. Seen in: P. Shepard and D. McKinley (eds.). 1969. The Subversive Science, Essays toward an Ecology of Man. Houghton Mifflin Co., Boston, MA. Pp. 230–241.

Woodwell, G. M. and T. G. Marples. 1968. Production and decay of litter and humus in an oak-pine forest and the influence of chronic gamma irradiation. Ecology 49:456–465.

Woodwell, G. M. and F. T. Martin. 1964. Persistence of DDT in soils of heavily sprayed forest stands. Science 145:481–483.

Woodwell, G. M. and L. N. Miller. 1963. Chronic gamma irradiation affects the distribution of radial increment in *Pinus rigida* stems. Science 139:222–223.

Woodwell, G. M. and J. K. Oosting. 1965. Effects of chronic gamma irradiation on the development of old field plant communities. Radiation Botany 5:205–222.

Woodwell, G. M. and E. V. Pecan (eds.). 1973. Carbon and the Biosphere. CONF 720510, Office of Information Services, Technical Information Center. U.S. Atomic Energy Commission, Washington, DC.

Woodwell, G. M. and A. L. Rebuck. 1967. Effects of chronic gamma radiation on the structure and diversity of an oak-pine forest. Ecological Monographs 37:53–69.

Woodwell, G. M. and H. H. Smith (eds.). 1969. Diversity and Stability in Ecological Systems. Brookhaven Symposia in Biology No. 22. Brookhaven National Laboratory, Upton, NY.

Woodwell, G. M. and A. H. Sparrow. 1963. Predicted and observed effects of chronic gamma irradiation on a near-climax forest ecosystem. Radiation Botany 3:231–237.

Woodwell, G. M. and A. H. Sparrow. 1965. Effects of ionizing radiation on ecological systems. Pp. 20–38. In: G. M. Woodwell (ed.). Ecological Effects of Nuclear War. US AEC Report BNL-917. Brookhaven National Laboratory, Upton, NY.

Woodwell, G. M. and R. H. Whittaker. 1968a. Effects of chronic gamma radiation on plant communities. Quarterly Review of Biology 43:42–55.

Woodwell, G. M. and R. H. Whittaker. 1968b. Primary production in terrestrial ecosystems. American Zoologist 8:19–30.

Woodwell, G. M., R. H. Whittaker, W. A. Reiners, G. E. Likens, C. C. Delwiche and D. B. Botkin. 1978. The biota and the world carbon budget. Science 199:141–146.

Woodwell, G. M., C. F. Wurster, Jr. and P. A. Isaacson. 1967. DDT residues in an East Coast estuary: A case study of biological concentration of a persistent insecticide. Science 156:821–824.

9

THE SAVANNAH RIVER ECOLOGY LABORATORY

Introduction

In 1951 the decision was made by the U.S. government to establish a second major plutonium production facility to complement Hanford, in Washington State, in order to support the development of the hydrogen bomb. A large area in South Carolina along the Savannah River was chosen due to the availability of land and access to water for cooling the nuclear reactors. Ultimately, the AEC acquired about 500 square kilometers (315 square miles) of primarily mixed forest and farmland. The facility was roughly circular in shape, with a border of 30 kilometers (18.6 miles) along the Savannah River (see maps in Reinig 1963, White and Gaines 2000). Approximately 6,000 people were relocated and hundreds of farms were abandoned to create what was called the Savannah River Site (SRS). The facility would include five nuclear production reactors, chemical separation and extraction plants, waste processing storage and disposal sites and hundreds of kilometers of roads. It was operated by the E. I. DuPont de Nemours Company as a contractor to the AEC with the primary goal to produce plutonium and tritium for use in hydrogen bombs.

The story of how E. P. Odum and the University of Georgia became associated with the SRS has been told many times (Hagan 1992, Craige 2001, Coleman and Camp 2011) and only a summary will be given here. Once planning for the SRS began, the AEC's Division of Biology and Medicine invited proposals for a pre-installation environmental inventory from the University of South Carolina, the University of Georgia and Ruth Patrick's group at the Academy of Natural Science in Philadelphia. Each institution was given a relatively small grant of $10,000 to do a component of the inventory. After the initial grant, only the University of Georgia stayed involved at the SRS, with E. P. Odum as the leader of the work. Odum saw a relationship with the AEC at the SRS as a potential opportunity for

DOI: 10.4324/9781003284901-11

long-term research with new tools such as radioisotopes for studying ecosystems. He had been influenced by Evelyn Hutchinson's work at Yale with radiotracers (see Chapter 2), partly through his brother Howard who was studying with Hutchinson for his Ph.D. degree. E. P. Odum himself had been trained in the pre–World War II descriptive ecology paradigm, having received his Ph.D. from the University of Illinois in 1939, but he longed to modernize the field with new ideas, including the ecosystem concept. Towards this goal he had been preparing a textbook on Ecology which would combine new concepts of energy flow, biogeochemical cycling and the ecosystem with the older, established concepts of the population, community and habitat. This was a brand new approach that became the standard for the post–World War II discipline of Ecology for many years. The first edition of his *Fundamentals of Ecology* came out in 1953, not long after Odum's first grant from the AEC in 1951. The textbook went through five editions over the next 50 years and it helped establish E. P. Odum as one of the most influential ecologists of the twentieth century.

The facilities of the SRS and the financial support from the AEC became critical to E. P. Odum's ambition to build the field of Ecology in general and, more specifically, to build Ecology's role as a component of his academic home at the University of Georgia. He proposed studying ecological succession at the SRS to take advantage of the thousands of hectares of farmland that had been abandoned when the facility was set up. This established a framework for long-term studies that he would use to understand how ecosystems develop. In this regard he advocated for functional studies, such as energy flow and mineral cycling, as opposed to taxonomic studies that had been well established for what was known as old-field succession. He wanted to use the new tools of radionuclide tracers but he had no background in radiation techniques. Undeterred, he embarked on a path to learn as rapidly as possible. His record of benchmarks in Radioecology is impressive: he became a consultant in Ecology to the Oak Ridge Radioecology Program and learned techniques from them while providing his knowledge of the ecosystem concept, he collaborated with his brother on a study of the coral reefs at Eniwetok that was supported by the AEC (see Chapter 3), he became a delegate to the 1955 Atoms for Peace Conference in Geneva, Switzerland, and during the years 1957–1958 he had a National Science Foundation Fellowship that allowed him to spend time studying Radioecology at the NTS, the Hanford Laboratory, in Washington State, and in Oxford, England. During this fellowship he wrote a chapter on radiation ecology that he included in the second edition of his textbook, published in 1959. Thus, by 1960 E. P. Odum was a national leader in the field of Radioecology!

Old Field Ecology at the SRS

Old fields, dominated by forbs and grasses, are an ephemeral ecosystem making up the early stages of succession on abandoned farmland. They last only about 10–20 years before woody vegetation takes over the site. This kind of ecosystem is easy to study because diversity is relatively low and plant biomass is readily harvested from sample plots by clipping for aboveground parts and shallow excavation for belowground parts. Also, animals are easy to sample with traps or sweep nets. Odum was well aware of the nature of old fields, from having grown up in rural North Carolina and from his first job after graduate school, as a naturalist at a preserve in New York State. He immediately started a program to understand the functional ecology of old fields on the SRS site once funding from the AEC started in late 1951. While he developed the knowledge base to use the new tools of Radioecology, he started an intense study of the old fields from an energy flow perspective. Graduate students were hired to help with the work using the AEC funding to support them. He first laid out the pattern of breeding birds through succession in the vicinity of the University of Georgia campus (Johnston and Odum 1956); although this represented the Piedmont region of the state, it was similar to what could be expected for the nearby SRS, which was located in the upper Coastal Plain region. Odum's paper includes a profile of plant communities along the successional gradient from abandoned farmland through old field stages to forest vegetation on which the distribution of birds is overlain. This is an iconic image of plant succession that shows the place of old fields in the time line of succession and he included it in his textbook (Odum 1971).

Starting in Fall 1951 an inventory was made of 600 fields at the SRS in terms of cropping history, vegetation composition and soil type. From this inventory, reference fields were identified and several were chosen for intensive study starting in the next growing season. These were monitored for biomass of living and dead "standing crop" or biomass. This data was used to establish net primary production, which was about 300 grams dry weight/m2/year. The regular pattern of seasonal production of live biomass followed by decomposition of dead biomass or litter was evident. It provided a record of the functioning of the overall ecosystem that contrasted with old field studies by other researchers who then mostly studied changes in species composition over time.

By the late 1950s, Odum and his coworkers were ready to start a program of field experiments with radioisotope tracers in the old fields that would last more than a decade. He had developed sufficient background in methods from other Radioecology groups to start experimenting with P-32. Over time his group would develop their own new methods (Odum et al. 1962a, Wiegert and Lindeborg 1964, Monk 1967 and use other radioisotopes, such as Zn-65 and Ca-45. Their initial approach was to label particular plant species in quadrats of old field vegetation by adding the radioactive P-32 directly to the individual plants. Animals in the quadrats would be collected at regular intervals and tested for radioactivity from the beta emissions of P-32 once the plants had been labeled. Any animal that emitted radioactivity must have been dependent on the plant that had been originally labeled, either directly or indirectly, for food. The data were plotted by activity density in terms of counts per minute/dry weight of material versus time. The time pattern of the curves for different species would show their relative trophic position along the food chain that started with the labeled plant species. Thus, herbivores would have an early peak in activity density while predators would have a later peak. In this way food chains in the old field could be isolated and food webs could be "untangled" (Odum and Kuenzler 1963) when different plant species were labeled in different quadrats of the same field. This was state-of-the-art functional ecology since previously it had not been possible to know definitively who was eating who in these ecosystems. An example of this kind of experiment is shown in Figure 9.1 for an old field at the SRS.

These kinds of labeled plots were termed "hot quadrats" due to their radioactivity (Odum 1965). In the study done by Wiegert et al. (1967), the actual quadrats were 10 meters by 20 meters in area and they were each labeled with approximately 4.5 millicuries of P-32. The results of these kinds of experiments were often surprising, as for example illustrated by de la Cruz and Wiegert (1967), who showed that a dominant ant species in the old field that had been thought to be a predator-scavenger was actually getting most of its energy by tending aphids and eating the honeydew that aphids produced. Because of the short half-life of P-32, these kinds of experiments could be followed for only about two months before the radioactivity dropped below detection limits.

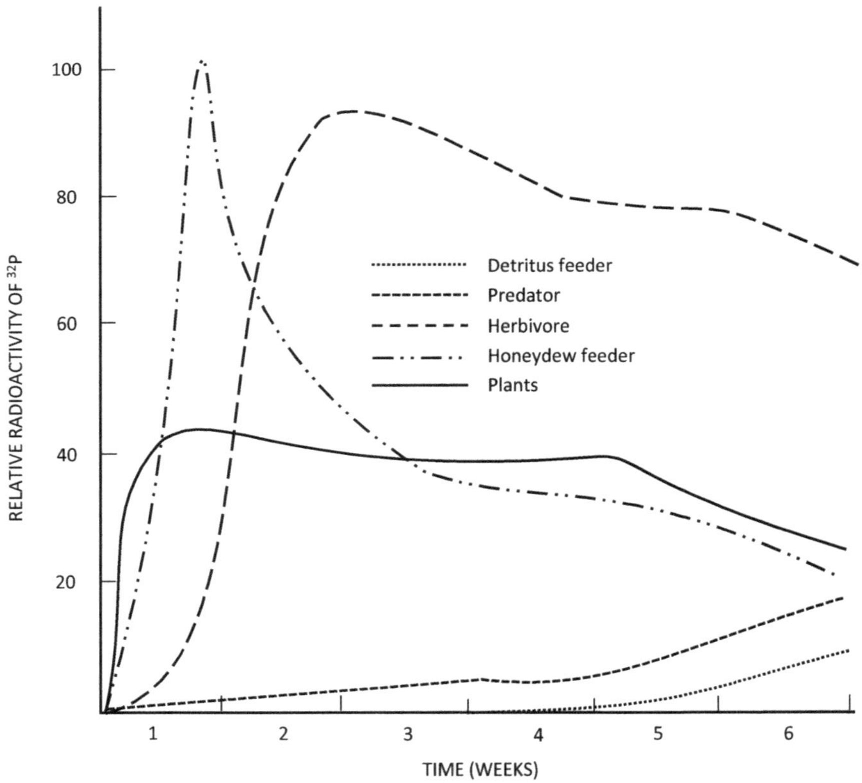

FIGURE 9.1 Conversion and utilization of biomass through trophic positions of an old field food chain as traced with the radioisotope P-32 (Wiegert et al. 1967).

Source: Used with permission of the Ecological Society of America.

Energy flow studies were also carried out in the old fields, both for individual species and whole ecosystems. These studies took the form of energy budgets, including production, respiration and other components. The First Law of Thermodynamics (the conservation law) was used to help balance the energy budgets with the inputs equaling the outputs plus change in storage. For the plants, gross production was divided into respiration and net production. For an animal, ingestion was divided into assimilation plus unassimilated food and, in turn, assimilation was divided into respiration and production. For example, Odum et al. (1962b) gave energetics data for three common animals in the old fields. They found winter resident savannah sparrows (*Passerculus sandwichensis*) had a production of 3.6 kcal/m2/year, the old field mice (*Peromyscus polionotus*) had a production of 6.7 kcal/m2/year and grasshoppers (Orthoptera) had a production of 25.6 kcal/m2/year. Also, Golley and Gentry (1964) studied the southern harvester ant (*Pogonmyrex badius*) and found its energy flow to range from 14 to 48 kcal/m2/year, which was higher than both of the vertebrates studied by Odum et al. (1962b). Menhinick (1967) estimated the energy flow for whole fields dominated by lespedeza (*Lespedeza cuneate*) with a diverse set of arthropods, which showed that the animal consumers utilized less than 1% of the total net primary production of 2553 kcal/m2/year. These energetics studies did not rely on radiotracer techniques, though some later work related animal respiration rates to excretion rates of certain isotopes. In addition to the basic research on functional ecology of the old fields at the SRS, Odum and his students carried out parallel studies in the Georgia salt marshes at Sapelo Island without AEC support (see, for example, Marples (1966) and Odum and Smalley (1959)).

It is important to note that all of the old field work done at the SRS was basic or academic in nature, rather than being directly applied to problems such as fallout or pollution; as was noted in Chapter 2, most of the early work on radiocesium and radiostrontium that entered agricultural food chains from fallout was done by agronomists rather than ecologists. Thus, the AEC seems to have been content to fund the University of Georgia's early work at the SRS to demonstrate the positive aspects of the new scientific technologies associated with radioisotopes rather than to focus on the negative aspects associated with fallout from nuclear testing. Only in the 1970s and afterwards did the University of Georgia researchers start to study radioactive pollution from the nuclear reactors at the SRS.

Establishment of the Savannah River Ecology Laboratory

After about a decade of annual grants from the AEC for ecological research by the University of Georgia at the SRS, the funding increased and became dedicated to the work being done there. On-site facilities were renovated to house the researchers and the laboratory was officially established in 1962. It was first known as the Laboratory of Radiation Ecology, but the name was later changed to the Savannah River Ecology Laboratory (SREL). Frank Golley, who had been hired to carry out research on old fields at the SRS after he graduated from Michigan State University in 1958, stepped into the role of director of the lab (Golley 2001). This started a new, expanded phase of Radioecology at the SRS with a series of new faculty hires including Richard Wiegert, Robert Beyers and Frank McCormick. Both Golley and Wiegert had carried out old field ecology research for their graduate work in Michigan and they quickly added to the line of work that E. P. Odum had started in 1951. For example, Golley did a major survey of the broomsedge (*Andropogon virginica*) stage of succession (Golley 1965) and Wiegert added to work on production and decomposition in the fields (Wiegert 1974, Wiegert and McGinnis 1975). Old field research continued during Golley's tenure as director of the SREL, but the work also diversified into new areas.

A major new initiative was irradiation of organisms and ecosystems with a portable Cs-137 device, termed the "Irradiator", which Golley and McCormick helped design (McCormick and Golley 1966). The CsCl source of 9,200 curies was obtained from the Oak Ridge National Laboratory and it was encapsulated in a movable lead shielding mechanism. Some earlier work by Georgia scientists at the SRS had investigated radiosensitivity of birds with a 100 curie Co-60 source, but the Irradiator allowed more opportunities for experimentation. After initial testing, the Irradiator was used in 18 different experiments, carried out by 10 different researchers, in 1964 and 1965 (Golley and McCormick 1966). McCormick (1969) used it to irradiate a 12-year-old pine forest for 200 hours in a test of acute system radiosensitivity. The forest was found to be highly sensitive to radiation compared to findings from studies published on other plant communities, and the results verified the research from the Brookhaven Laboratory (see Chapter 8). Cumulative exposures of 3000 rads or more (within 10 meters of the source) were lethal to both slash pines and long-leaf pines (*Pinus* spp.), and exposures of 400–500 rads (from 10 to 30 meters from the source) produced severe growth inhibition. The microclimate changed when the trees died and conditions resembled what would be found in an old field rather than a forest in terms of temperature and light penetration. After irradiation had ceased, old field forbs and grasses colonized the site. Also, Carl Monk, who had succeeded McCormick when he moved to the University of North Carolina in 1965, irradiated an old field community for 400 hours (Monk 1966). In the highest-exposure zone of 160 krads, all of the original vegetation was killed but no significant zonation of effects beyond this was seen. After irradiation ceased, a sedge, *Fimbristylis autumnalis*, which was not present before irradiation, invaded what had been the highest-exposure zone. The success of this species was hypothesized to have been due to "a combination of radiation resistance and sensitivity to competition". Finally, Golley and his coworkers conducted a variety of irradiation experiments with mammals found on the SRS. This included multiple studies with rodents but also with larger mammals including bobcats (*Lynx rufus*) (Golley et al. 1965). It must have been a challenge to place a live bobcat into the irradiation chamber since, according to Golley's (1966) own textbook on South Carolina mammals, this was the largest carnivore found in the state!

A major side-project occurred in 1966 when Golley and a team of Georgia faculty and graduate students joined a large project to study the feasibility of creating a sea-level canal in Panama or Colombia that might be excavated with nuclear explosives. This was part of the AEC's Plowshare Program to develop peaceful uses for thermonuclear explosives. The concept had first been outlined by physicists at the Lawrence Radiation Lab in 1957, and by 1961 the AEC had formed a Division of Peaceful Nuclear Explosives to pursue the concept. Potential uses were the excavation of canals, harbors and reservoirs and recovery of mineral deposits (see Chapter 5 for the Chariot Project in Alaska as an example). Creation of a new sea-level transisthmian, interoceanic canal was the most ambitious objective in the Plowshare Program, which had been championed by Edward Teller (Findlay 1990, Kirsch 2005). The existing canal uses a series of locks to move ships over the mountains of Panama in order to connect the Atlantic and Pacific Oceans, but the proposed sea-level canal would do so more cheaply and efficiently, according to Plowshare advocates. Overall, this was a fantastically ambitious undertaking that would have required "anywhere from 260 to more than 700 nuclear devices, producing a total energy release in the range of 20,000 times as much as the bomb that destroyed Hiroshima" (Regis 2015). The feasibility study was led by the Battelle Memorial Institute, a large consulting firm based in Columbus, Ohio. The AEC funded a large number of ecological studies as part of the work (Martin 1969). Golley's team spent two years studying the rainforests of Colombia and Panama, and their work was published in a book focusing on mineral cycling (Golley et al. 1975). The work included data on forest biomass and chemical content along with measurements of hydrologic flows and mineral cycling processes. A computer model of the potassium cycle of the forest, which was simulated on an analog computer, is described in the book, and a model of the magnesium cycle was analyzed by Child and Shugart (1972). Jordan (2001) credited the work as "the first ecosystem study in the Tropics in which major structures such as biomass of trees and functions such as rates of nutrient transfer were quantified" and Paul Richards (1975) in the foreword of the book called it "a notable contribution to understanding the tropical forest as an ecosystem". Ultimately the sea-level canal project was found to have too much risk of radioactive contamination, among other problems, and it was canceled. However, the work of Golley and his team, along with some of the other ecological research done in the feasibility study, was a valuable contribution of the idea of nuclear excavation of a sea-level canal, which in hindsight appears more than a bit delusional.

Golley moved back to the University of Georgia campus in Athens in 1967 to direct the newly formed Institute of Ecology and Robert Beyers took his position as director of the SREL. Beyers had gotten his doctorate degree studying with H. T. Odum on the use of aquatic microcosms for ecological research (Beyers 1963, 1964). He continued to work with small, laboratory-scale systems at the SREL where, among other activities, he developed a standard microcosm that could be used for many kinds of studies (Beyers and Odum 1993). The microcosm was originally derived from a sewage oxidation pond and it was cultured in 250 ml Erlenmeyer glass flasks in a growth chamber. A stock culture was maintained and inoculate from it was used to start an experiment. The biota included three species of algae, protozoans, an ostracod and several species of bacteria. If the initial culture media contained a nitrogen source, then an autotrophic succession would be stimulated (Cooke 1967) with higher production than respiration. If the initial culture media contained an organic compound, then a heterotrophic succession would be stimulated (Gordon et al. 1969) with higher respiration than production. However, over the long term (which for the microcosm was several months) either system would eventually reach a state of balanced metabolism where production equaled respiration. The successional patterns of these microcosms in terms of metabolism were of special interest to E. P. Odum, who was developing theories of ecosystem development that had started with his early research on old fields. Odum highlighted Beyer's microcosm in the third edition of his *Fundamentals of Ecology* text, especially with a graph that compared succession of a forest with a microcosm, showing similar patterns but with different time scales. Thus, the microcosm confirmed and reinforced his thinking on how ecosystem structure and function change over time and it figured prominently in his classic paper on ecosystem development (Odum 1969).

In a direct radioecology application, Ferens and Beyers (1972) used the standard microcosm in an irradiation experiment. Inoculate samples were irradiated at 10^5 rads and 10^6 rads, then seeded into the flasks and tracked over a 40-day period, relative to a control treatment of microcosms that was not irradiated. Data on structure and function of the microcosms in the experiment are shown in Table 9.1.

Both irradiated treatments had depressed values of ecosystem performance due to the radiation stress but, surprisingly, the treatment receiving the higher levels of radiation (10^6 rads) was affected less than the treatment receiving lower levels of radiation (10^5 rads). The explanation for this seemingly contradictory result was that "the community appeared to adapt to the stress through changes in species abundance" (Ferens and Beyers 1972). In other words, the system self-organized in response to the stress and the set of microcosms that received more radiation maintained its ecosystem function to a greater degree than the set that received less radiation. In terms of the algae species in the microcosms, the normally dominant green alga *Chlorella*, was more sensitive to irradiation than the blue-green alga *Schizothrix*, which grew better without competition and replaced *Chlorella* as the dominant primary producer in the irradiated microcosms.

TABLE 9.1 Data on ecosystem performance from the irradiation of aquatic microcosms (Ferens and Beyers 1972). Values of the two irradiated treatments are reported as percentages of a replicate set of control microcosms that were not irradiated and they were interpolated from graphs given in the manuscript.

Parameter	*10^5 rads*	*10^6 rads*
Biomass	50	80
Chlorophyll a	40	50
Gross production	60	70

Cooling Water and Thermal Pollution

Aquatic ecology increased in importance at the SREL during Beyers' time as director (1967–1972) and this facilitated the transition into studies of thermal pollution. With the growth of the commercial nuclear power industry in the U.S., thermal pollution of water used to cool reactors increased as an environmental issue. The reactors needed to be operated at extremely high temperature in order to support the nuclear reactions. The heat from the reactor is dispersed into water that is circulated through the system. Thermal pollution occurs when the heated water is released into the environment after having been used to cool the reactors. Moderate increases in water temperature can have positive effects, such as increased productivity and growth of aquatic organisms, but negative effects dominate with higher temperature, such as mortality of organisms that can't escape the heat through movements. Studies of thermal pollution effects increased through the 1960s, and by the 1970s the ecologists at the SREL became had become leaders in the research.

The long-term operation of the nuclear reactors at the SRS caused changes in aquatic ecosystems of Par Pond, the artificial impoundment that was constructed to hold on-site cooling water, and in the tributaries of the Savannah River that received heated water discharges. For examples, the alligators stayed active longer in the winter (Murphy and Brisbin 1974) and turtles grew larger (Christy et al. 1974) in the heated portions of Par Pond, while canopy trees were killed in swamps bordering streams receiving thermal discharges (Sharitz et al. 1974).

The SREL created a research program for thermal ecology studies in 1970 and utilized the environments of the SRS as a field laboratory over time (Gibbons and Sharitz 1981). To some extent this was a move away from academic-oriented research that had characterized the early Radioecology at the SRS, such as using tracers to identify food chains in the old fields, into a new, related area of applied research associated with the growing nuclear power industry. Radioecology was undergoing a transition in the early 1970s (as will be discussed in Chapter 10) and this shift to studies of thermal ecology was an opportunistic reaction for the SREL scientists. Many examples of self-organization of wetlands have been described from the SRS, with changes caused by flooding and heated discharge waters (Sharitz et al. 1974). Gibbons and Sharitz (1974) provided a photographic record of one ecosystem showing mist from condensation of hot water eerily rising from a swamp, stands of dead trees and marsh plants whose growth was stunted by thermal pollution. Reviews of the early thermal ecology research are given in the two national symposia on the subject that were hosted at the SREL and by Kingsland (2019).

Heritage

The last nuclear reactor was shut down in 1993 at the SRS, but radioactive contamination is a legacy of years of production of plutonium and tritium for nuclear weapons. The site has become a repository of radioactive wastes from other locations, which further complicates the problems. It is now operated for the U.S. Department of

Energy, a descendent of the AEC, by Savannah River Nuclear Solutions LLC. The University of Georgia's SREL continues to be supported by the U.S. Department of Energy after more than 70 years. The focus areas of ecological research at the SRS have changed over the decades, and at present much work is being done on radioactive pollution of the site. Contamination has been identified in a variety of habitats (Whicker et al. 1990, Pinder et al. 1990, Leaphart et al. 2020), and wildlife from the site are providing useful information about long-term effects of pollution from radionuclides (Kennamer et al. 2017, Lamb et al. 1991, Gaines et al. 2021). Thermal pollution studies declined as the reactors were shut down, but this form of pollution also has a legacy of impacts which continue to be studied (Martin et al. 1977).

A more positive legacy is that the SRS remains "the largest protected natural area in the eastern United States" (Cohn 1994) in part because the contamination requires that access to the site be limited for security and safety. The SRS was formally designated a National Environmental Research Park in 1972 by the AEC, which ensures that SREL personnel and other scientists may be able to study the spectrum of habitats that are represented on the site. Also, the old fields that dominated much of the area when the SRS was established in 1951 have been transformed through succession into forest that is managed by the U.S. Forest Service for pulpwood and lumber production (White and Gaines 2000, Kilgo and Blake 2005).

A number of University of Georgia faculty who made significant contributions to the field of Ecology started their professional careers at the SREL during the Cold War era and they represent an important part of the heritage of the lab. Prime among this group was E. P. Odum, who wrote the first grant proposal to the AEC for ecological research at the SRS. Odum became one of the leading ecologists of the twentieth century by emphasizing the ecosystem approach (Likens 2001, Craige 2001, Barrett and Likens 2002). He built an academic empire at the University of Georgia in the form of the Institute of Ecology (Barrett and Barrett 2001) and he helped educate a generation with his textbook *Fundamentals of Ecology* and his other books. Odum was elected president of the Ecological Society of America (ESA) in 1964 (Anonymous 1964) and he received many honors throughout his career. Although a leading figure in Ecology for more than 50 years, as noted by Egerton (2015), before his first contract with the AEC in 1951 he was "a small fish in a small pond" at the University of Georgia in the 1940s. Certainly AEC support of Odum's early research at the SRS was critical to his development as an ecologist. Hagan (1992) suggested:

> What began as a small and uncertain source of support rapidly developed into a long-term relationship between the University of Georgia and the AEC. By 1960 the annual budget for ecological research at Savannah River had increased to $60,000, and the AEC had agreed to establish a permanent ecological laboratory at the site. . . . One can only speculate about how successful Eugene Odum's research program would have been without the long-term support of the AEC. It seems safe to say that the University of Georgia would not have become such a major center of ecological research were it not for atomic energy.

Through the 1960s and 1970s Odum broadened his research funding and his perspective beyond Radioecology, but the early support from the AEC was important. His acclaimed paper on ecosystem development (Odum 1969) shows both his ties to the early work on functional ecology of succession at the SRS and his later broadening perspective beyond Radioecology. The heart of the paper is a "tabular model of ecological succession" comparing developmental stages of succession with mature stages. It includes 24 ecosystem attributes grouped under the following headings: community energetics, community structure, life history, nutrient cycling, selection pressure and overall homeostasis. The model covers all the major topics of Ecology, as does his textbook, and it is a kind of road map for successional change in Nature. It remains a benchmark paper in synthesis that has been reprinted many times (see, for example, Golley 1977, Real and Brown 1991). It was controversial when it was published and it still continues to elicit review and debate (Kominoski et al. 2018). The influences from succession research done at the SREL in the 1950s and 1960s on old fields and microcosms are evident, especially in the paper's first figure, which plots production, respiration and biomass over successional time scales. The paper also discusses the relevance of ecosystem development to human ecology, and the subtitle is "an understanding of ecological succession provides a basis for resolving man's conflict with nature". The themes of the paper have been taken up by human ecologists (see, for example, Benyus 1997, Wann 1990), demonstrating the paper's degree of influence. Odum later expanded the tabular model to theorize how stress affects ecosystem development (Odum 1985). In a sense, Odum's strategy of ecosystem development paper was a philosophical jump beyond the pre–World War II debates about succession and superorganisms. In the present day a further philosophical jump has been made with the Panarchy paradigm (Carpenter and Turner 2001) which suggests that systems cycle through four characteristic phases: 1) rapid growth and exploitation, 2) conservation, 3) collapse or release, and 4) renewal or reorganization.

Led by Frank Golley, other SREL scientists left the lab to move into faculty positions on the Athens, Georgia, campus where they made important contributions to the field of Ecology. Golley left in 1967 to become director of the newly formed Institute of Ecology. He remained prolific as a scientist while carrying out his administrative duties and he published and edited a number of books on Ecology and Environmental Science. For the purposes of this treatment, one of his most important books was a review of the ecosystem concept in Ecology (Golley 1993). Since he was a participant in a good deal of this history, his book is an excellent combination of comprehensive literature review with his personal perspectives. Golley was elected president of the Ecological Society of America in 1976 (Odum 1976) and his productive career as a scientist has been celebrated multiple times (Jordan 2006, Pomeroy et al. 2007, Covich 2015). Several other faculty members used their early research in Cold War–era Radioecology done at the SREL as a boost in their later careers. Richard Wiegert continued work on ecological energetics and he later focused on computer simulation of ecosystems (Wiegert 1975) with detailed efforts on salt marshes and the thermal spring in Yellowstone

National Park in Wyoming. Carl Monk continued his work on plant ecology, becoming a leading authority on southeastern U.S. vegetation (Monk 1971). One person who stayed at the SREL for her whole career was Rebecca Sharitz. She built on her thermal ecology work in hardwood swamps at the SRS to become a nationally known expert in wetland ecology (Batzer and Sharitz 2006).

Lastly, in terms of heritage, the SREL has directly helped educate a number of students who worked there, since it is a part of a university. Starting with the first three graduate students who worked on E. P. Odum's initial grant proposal, many students were supported over the years by AEC funding. An example is Michael Smith, who did his postdoctoral research at the SREL and later became the director after Robert Beyers left. Additionally, many undergraduates got research experience at the SREL during the summers through a special grant program funded by the AEC, including this author's brother (Kangas and Coleman 1970). The lab's annual reports record student participation at the site, which over time has involved hundreds of individuals.

Literature Cited

Anonymous. 1964. President Eugene P. Odum. Bulletin of the Ecological Society of America 45:129–130.

Barrett, G. W. and T. L. Barrett (eds.). 2001. Holistic Science: The Evolution of the Georgia Institute of Ecology (1940–2000). Taylor & Francis, New York, NY.

Barrett, G. W. and G. E. Likens. 2002. Eugene P. Odum: Pioneer of ecosystem ecology. Bioscience 52:1047–1048.

Batzer, D. P. and R. R. Sharitz (eds.). 2006. Ecology of Freshwater and Estuarine Wetlands. University of California Press, Berkeley, CA.

Benyus, J. M. 1997. Biomimicry, Innovation Inspired by Nature. William Morrow, New York, NY.

Beyers, R. J. 1963. The metabolism of twelve aquatic laboratory microecosystems. Ecological Monographs 33:281–306.

Beyers, R. J. 1964. The microcosm approach to ecosystem biology. American Biology Teacher 26:491–498.

Beyers, R. J. and H. T. Odum. 1993. Ecological Microcosms. Springer-Verlag, New York, NY.

Carpenter, S. R. and M. G. Turner. 2001. Panarchy 101. Ecosystems 4:389.

Child, G. I. and H. H. Shugart, Jr. 1972. Frequency response analysis of magnesium cycling in a tropical forest ecosystem. Pp. 103–135. In: B. C. Patten (ed.). Systems Analysis and Simulation in Ecology, Volume 2. Academic Press, New York, NY.

Christy, E. J., J. O. Farlow, J. E. Bourque and J. W. Gibbons. 1974. Enhanced growth and increased body size of turtles living in thermal and post-thermal aquatic systems. Pp. 277–284. In: J. W. Gibbons and R. R. Sharitz (eds.). Thermal Ecology. CONF-730505, U.S. Atomic Energy Commission, Washington, DC.

Cohn, J. P. 1994. A national natural laboratory. Bioscience 44:727–730.

Coleman, D. C. and T. Camp. 2011. The Prehistory of the Odum School of Ecology. Odum School of Ecology, University of Georgia, Athens, GA.

Cooke, G. D. 1967. The pattern of autotrophic succession in laboratory microcosms. Bioscience 17:717–721.

Covich, A. P. 2015. On Frank Golley's international and interdisciplinary insights for a Twenty-First Century earth stewardship based on environmental ethics. Pp. 431–450. In: R. Rozzi, et al. (ed.). Earth Stewardship, Ecology and Ethics 2. Springer International Publ., Switzerland.

Craige, B. J. 2001. Eugene Odum, Ecosystem Ecologist & Environmentalist. University of Georgia Press, Athens, GA.

de la Cruz, A. A. and R. G. Wiegert. 1967. 32-Phosphorus tracer studies of a horseweed-aphid-ant food chain. American Midland Naturalist 77:501–509.

Egerton, F. N. 2015. A Centennial History of the Ecological Society of America. CRC Press, Boca Raton, FL.

Ferens, M. C. and R. J. Beyers. 1972. Studies of a simple laboratory microecosystem: Effects of stress. Ecology 53:709–713.

Findlay, T. 1990. Nuclear dynamite: The Peaceful Nuclear Explosions Fiasco. Pergamon Press Australia. Rushcutters Bay, Australia.

Gaines, K. F., P. M. Novak and J. M. Novak. 2021. Ecological half-life of radiocesium in white-tailed deer on the Department of Energy's Savannah River Site: What can a half century of field monitoring tell us? Journal of Environmental Radioactivity 235–236(2021):10654.

Gibbons, J. W. and R. R. Sharitz. 1974. Thermal alteration of aquatic ecosystems. American Scientist 62:660–670.

Gibbons, J. W. and R. R. Sharitz. 1981. Thermal ecology: Environmental teachings of a nuclear reactor site. Bioscience 31:293–298.

Golley, F. B. 1965. Structure and function of an old-field broomsedge community. Ecological Monographs 35:113–137.

Golley, F. B. 1966. South Carolina Mammals. The Charleston Museum. Charleston, SC.

Golley, F. B. (ed.). 1977. Ecological Succession, Benchmark Papers in Ecology, Volume 5. Dowden, Hutchingon & Ross, Inc., Stroudsburg, PA.

Golley, F. B. 1993. A History of the Ecosystem Concept in Ecology, More Than the Sum of the Parts. Yale University Press, New Haven, CT.

Golley, F. B. 2001. Establishing the network. Pp. 38–67. In: G. W. Barrett and T. L. Barrett (eds.). Holistic Science: The Evolution of the Georgia Institute of Ecology (1940–2000). Taylor & Francis, New York, NY.

Golley, F. B. and J. B. Gentry. 1964. Bioenergetics of the southern Harvester ant, *Pogonomyrmex badius*. Ecology 45:217–225.

Golley, F. B. and J. F. McCormick. 1966. Development of a 137-Cesium facility for ecological research. TID-22499, U.S. Atomic Energy Commission, Washington, DC.

Golley, F. B., J. T. McGinnis, R. G. Clements, G. I. Child and M. J. Duever. 1975. Mineral Cycling in a Tropical Moist Forest Ecosystem. University of Georgia Press, Athens, GA.

Golley, F. B., E. L. Rauber, E. L. Morgan and J. H. Jenkins. 1965. Effect of acute gamma radiation on wild opossum, gray fox, raccoon and bobcat. Health Physics 11:1573–1576.

Gordon, R. W., R. J. Beyers, E. P. Odum and R. G. Eagon. 1969. Studies of a simple laboratory microecosystem: Bacterial activities in a heterotrophic succession. Ecology 50:86–100.

Hagan, J. B. 1992. An Entangled Bank: The Origins of Ecosystem Ecology. Rutgers University Press, New Brunswick, NJ.

Johnston, D. W. and E. P. Odum. 1956. Breeding bird populations in relation to plant succession on the piedmont of Georgia. Ecology 37:50–62.

Jordan, C. F. 2001. Tropical ecology with the Institute of Ecology. Pp. 261–276. In: G. W. Barrett and T. L. Barrett (eds.). Holistic Science: The Evolution of the Georgia Institute of Ecology (1940–2000). Taylor & Francis, New York, NY.

Jordan, C. F. 2006. The life and career of Frank B. Golley (1930–2006) and his role in awakening North American ecology to the international realm. Tropical Ecology 47:155–157.

Kangas, M. and D. C. Coleman. 1970. A study of five techniques for the estimation of numbers and activity of bacteria in soil. Pp. 310–317. In: 1970 Annual Report, Part 2. Savannah River Ecology Laboratory, University of Georgia, Aiken, SC.

Kennamer, R. A., R. E. Oldenkamp, J. C. Leaphart, J. D. King, A. L. Bryan, Jr. and J. C. Beasley. 2017. Radiocesium in migratory aquatic game birds using contaminated U.S. Department of Energy reactor-cooling reservoirs: A long-term perspective. Journal of Environmental Radioactivity 171:189–199.

Kilgo, J. C. and J. I. Blake (eds.). 2005. Ecology and Management of a Forested Landscape, Fifty Years on the Savannah River Site. Island Press, Washington, DC.

Kingsland, S. 2019. Fifty years ago: The development of thermal ecology at the Savannah River Ecology Laboratory. Bulletin of the Ecological Society of America 100:1–3. (Article e01518).

Kirsch, S. 2005. Proving Grounds, Project Plowshare and the Unrealized Dream of Nuclear Earthmoving. Rutgers University Press, New Brunswick, NJ.

Kominoski, J. S., E. E. Gaiser and S. G. Baer. 2018. Advancing theories of ecosystem development through long-term ecological research. Bioscience 68:554–562.

Lamb, T., J. W. Bickham, J. W. Gibbons, M. J. Smolen and S. McDowell. 1991. Genetic damage in a population of slider turtles (*Trachemys scripta*) inhabiting a radioactive reservoir. Archives of Environmental Contamination and Toxicology 20:138–142.

Leaphart, J. C., A. M. Korotasz, A. L. Bryan, Jr. and J. C. Beasley. 2020. Environmental fate of radiocesium in biota inhabiting a contaminated ecosystem on the U.S. Department of Energy's Savannah River Site. Journal of Environmental Radioactivity 222:106321.

Likens, G. E. 2001. Eugene P. Odum, the ecosystem approach and the future. Pp. 307–328. In: G. W. Barrett and T. L. Barrett (eds.). Holistic Science: The Evolution of the Georgia Institute of Ecology (1940–2000). University of Georgia Press, Athens, GA.

Marples, T. G. 1966. A radionuclide tracer study of arthropod food chains in a *Spartina* salt marsh ecosystem. Ecology 47:270–277.

Martin, C. E., E. J. Christy and K. W. McLeod. 1977. Changes in the vegetation of a South Carolina swamp following cessation of thermal pollution. Journal of the Elisha Mitchell Scientific Society 93:173–176.

Martin, W. E. 1969. Radioecology and the feasibility of nuclear canal excavation. Pp. 9–22. In: D. J. Nelson and F. C. Evans (eds.). Symposium on Radioecology, Proceedings of the Second National Symposium, Ann Arbor, Michigan. CONF-670503, U.S. Atomic Energy Commission, Washington, DC.

McCormick, J. F. 1969. Effects of ionizing radiation on a pine forest. Pp. 78–87. In: D. J. Nelson and F. C. Evans (eds.). Symposium on Radioecology, Proceedings of the Second National Symposium, Ann Arbor, Michigan. CONF-670503, U.S. Atomic Energy Commission, Washington, DC.

McCormick, J. F. and F. B. Golley. 1966. Irradiation of natural vegetation, an experimental facility, procedures and dosimetry. Health Physics 12:1467–1474.

Menhinick, E. F. 1967. Structure, stability, and energy flow of plants and arthropods in a *Sericea lespedeza* stand. Ecological Monographs 37:255–272.

Monk, C. D. 1966. Effects of short-term gamma radiation on an old field. Radiation Botany 6:329–335.

Monk, C. D. 1967. Radioisotope tagging through seed soaking. Bulletin of the Georgia Academy of Science 25:13–17.

Monk, C. D. 1971. Species and area relationship in the eastern deciduous forest. Journal of the Elisa Mitchell Scientific Society 87:227–230.

Murphy, T. M., Jr. and I. L. Brisbin, Jr. 1974. Distribution of alligators in response to thermal gradients in a reactor cooling reservoir. Pp. 313–321. In: J. W. Gibbons and R. R. Sharitz (eds.). Thermal Ecology. CONF-730505 U.S. Atomic Energy Commission, Washington, DC.

Odum, E. P. 1960. Organic production and turnover in old field succession. Ecology 41:34–49.

Odum, E. P. 1965. Feedback between radiation ecology and general ecology. Health Physics 11:1257–1262.

Odum, E. P. 1969. The strategy of ecosystem development. Science 164:262–270.

Odum, E. P. 1971. Fundamentals of Ecology, Third Edition. W. B. Saunders Co., Philadelphia, PA.

Odum, E. P. 1976. Frank Golley, President 1976–1977. Bulletin of the Ecological Society of America 57:7.

Odum, E. P. 1985. Trends expected in stressed ecosystems. Bioscience 35:419–422.

Odum, E. P., C. E. Connell and L. B. Davenport. 1962b. Population energy flow of three primary consumer components of old-field ecosystems. Ecology 43:88–96.

Odum, E. P. and E. J. Kuenzler. 1963. Experimental isolation of food-chains in an old-field ecosystem with the use of phosphorus-32. Pp. 113–120. In: V. Schultz and A. W. Klement, Jr. (eds.). Radioecology, Proceedings of the First National Symposium held at Colorado State University, Fort Collins, CO. Reinhold Publishing Corporation, New York, NY.

Odum, E. P., R. P. Martin and B. C. Loughman. 1962a. Scanning systems for the rapid determination of radioactivity in ecological materials. Ecology 43:171–173.

Odum, E. P. and A. E. Smalley. 1959. Comparison of population energy flow of a herbivorous and a deposit-feeding invertebrate in a salt marsh ecosystem. Proceedings of the National Academy of Sciences 45:617–622.

Pinder, J. E., III, K. W. McLeod and D. C. Adriano. 1990. Atmospheric deposition, resuspension, and root uptake of Pu in corn and other grain-producing agroecosystems near a nuclear fuel facility. Health Physics 59:853–867.

Pomeroy, L. R., D. C. Coleman and D. A. Crossley, Jr. 2007. Resolution of respect, Frank B. Golley 1930–2006. Bulletin of the Ecological Society of America 88:15–17.

Real, L. A. and J. H. Brown (eds.). 1991. Foundations of Ecology, Classic Papers with Commentaries. University of Chicago Press, Chicago, IL.

Regis, E. 2015. Monsters: The Hindenburg Disaster and the Birth of Pathological Technology. Basic Books, New York, NY.

Reinig, W. C. 1963. The 1951 preoperational environmental survey for the Savannah River Plant: In retrospect. Health Physics 9:83–85.

Richards, P. 1975. Forward. Pp. xvii–xviii. In: Mineral Cycling in a Tropical Moist Forest Ecosystem. F. B. Golley, J. T. McGinnis, R. G. Clements, G. I. Child and M. J. Duever. University of Georgia Press, Athens, GA.

Sharitz, R. R., J. W. Gibbons and S. C. Gause. 1974. Impact of production-reactor effluents on vegetation in a southeastern swamp forest. Pp. 356–362. In: J. W. Gibbons and R. R. Sharitz (eds.). Thermal Ecology. CONF-730505 U.S. Atomic Energy Commission, Washington, DC.

Sharitz, R. R., J. E. Irwin and E. J. Christy. 1974. Vegetation of swamps receiving reactor effluents. Oikos 25:7–13.

Wann, D. 1990. Biologic, Environmental Protection by Design. Johnson Books, Boulder, CO.

Whicker, F. W., J. E. Pinder III, J. W. Bowling, J. J. Alberts and I. L. Brisbin, Jr. 1990. Distribution of long-lived radionuclides in an abandoned reactor cooling reservoir. Ecological Monographs 60:471–496.

White, D. L. and K. F. Gaines. 2000. The Savannah River Site: Site description, land use and management history. Studies in Avian Biology 21:8–17.

Wiegert, R. G. 1974. Litterbag studies of microarthropod populations in three South Carolina old fields. Ecology 55:94–102.
Wiegert, R. G. 1975. Simulation models of ecosystems. Annuals of Ecology and Systematics 6:311–338.
Wiegert, R. G. and R. G. Lindeborg. 1964. A "stem well" method of introducing radioisotopes into plants to study food chains. Ecology 45:406–410.
Wiegert, R. G. and J. T. McGinnis. 1975. Annual production and disappearance of detritus on three South Carolina old fields. Ecology 56:129–140.
Wiegert, R. G., E. P. Odum and J. H. Schnell. 1967. Forb-arthropod food chains in a one-year experimental field. Ecology 48:75–83.

PART III

10

RADIOECOLOGY SINCE THE COLD WAR

Changes at the AEC

Major changes took place in the early 1970s that affected the AEC, leading to a decline in its role as the dominant funding source for ecosystem studies. The Cold War, which had been the initial trigger for ecosystem studies, continued well into the late 1980s, but the nature of Radioecology research changed more than a decade before this time. There were three main causes for the decline in in AEC support for ecosystem research: 1) a declining perception in the threat of nuclear war, 2) an increasing need for competing ecological research associated with commercial nuclear power plants and 3) the retirement of John Wolfe as the guiding force for AEC-funded ecosystem ecology.

Geopolitically, the 1950s and 1960s had been a time of high anxiety about the possibility of nuclear war. Events such as the Soviets' triumphal launch of the satellite *Sputnik* in 1957, the construction of the Berlin Wall in 1961 and the Cuban missile crisis in 1962 created this growing anxiety in the U.S. However, once the East and the West had achieved the status of mutually assured destruction with the growth of nuclear arsenals, a kind of stalemate emerged in the Cold War and by the early 1970s, the American society became inured to a degree to the threat of nuclear war. Arms control efforts contributed to this feeling, first with the Limited Test Ban Treaty in 1963 and later with the Nuclear Non-Proliferation Treaty (NPT) in 1968. Although not all countries signed the NPT, the major nuclear powers did and this began a cultural shift in which fear of nuclear war began to diminish or at least move from the forefront of social concern. With less of a perceived threat of nuclear war, there was less motivation to fund ecological research associated with it.

The development of nuclear reactors for commercial power generation was a major cause for change in the AEC and in Radioecology as a discipline in

DOI: 10.4324/9781003284901-13

the U.S. The first commercial nuclear power plant came online in 1958 and there was great hope that this technology would grow and provide a cheap, relatively clean source of electrical energy. In the early 1970s an energy crisis caused by Middle East oil embargos provided further incentives for the growth of the nuclear power industry. By 1970 about 20 nuclear power plants had come online in the U.S. and the number doubled by the mid-1970s (Rolph 1979). The AEC was charged with overseeing the development of nuclear power, which proved to be a major administrative challenge. Nuclear power plants essentially utilize the same kinds of initial physical-chemical processes as a nuclear bomb but, of course, under controlled conditions. Safety of these reactors was and is of paramount importance, especially to the people living near the nuclear reactors and to the natural and managed ecosystems in the area surrounding the power plant. The National Environmental Policy Act, which came into force in 1970, required an environmental impact evaluation for federally regulated projects like nuclear power plants and the AEC was one of the first national agencies that had to deal with the new regulation. Initially, it had difficulty meeting this requirement (see, for example, Gillette 1971) and ultimately the AEC shifted its funding for ecological research in this direction. Some ecologists who had earlier been supported by the AEC in ecosystem studies moved to studying ecological aspects of nuclear power generation (see, for example, Auerbach 1971, Auerbach et al. 1976). This type of ecological research progressed so fast that within a decade textbooks were being published on the subject (Eichholz 1976, Glasstone and Jordan 1980). Overall, to accommodate changes in atomic energy, the AEC was split into two parts by the Energy Reorganization Act of 1974: the Nuclear Regulatory Commission, which dealt with regulation of nuclear power generation, and the Energy Research and Development Administration (ERDA), which dealt with atomic progress and environmental issues.

A final cause for the AEC's declining role in support of ecosystem studies was the retirement of John Wolfe in 1972. Wolfe has been called the "father" of Radioecology in the United States (Seymour 1976) because of his guidance of the AEC's ecology program from 1958, when he became chief of the Environmental Sciences Branch, until his retirement. Although he was trained in pre–World War II descriptive plant ecology, his major professor at Ohio State University was E. N. Transeau who had done one of the very first ecological energetics studies (Transeau 1926) and who may have planted a seed for Wolfe's later interest in ecosystems. Wolfe was praised by many of the AEC-supported project leaders for his research vision, which was outlined to some extent in his addresses to the national symposia on Radioecology that the AEC organized (Wolfe 1963, 1969). As an example of this recognition, Janice Beatley (1976) dedicated her book on the plant ecology of the Nevada Test Site as follows:

To
John N. Wolfe

ecologist, teacher, poet
in whose wisdom description and explanation
were the inseparable pursuits of ecology,
and who knew the large pleasures
in achievement of both,
this work is gratefully dedicated.

Intending to write a textbook on Radioecology, Wolfe took administrative leave from the AEC in 1970, but he developed health problems which led to his retirement in 1972 and ultimately his death in 1974 (Sprugel 1975). The loss of his voice in funding decisions may have contributed significantly to support for new directions in ecological research by the AEC in the mid-1970s.

The last major ecosystem project supported by the AEC was located in the temperate deciduous forest in northern Wisconsin. The research was carried out in collaboration with the U.S. Forest Service's North Central Forest Experiment Station, which provided land for the project. Planning for the work began in 1968 and the original goal was for a five-year chronic exposure of the forest to a gamma radiation source. Pre-irradiation background studies were conducted from 1969 to 1971. Irradiation of the forest occurred from May to October 1972, but "as a consequence of changing priorities in the AEC radioecological research program" the project was terminated in 1973 (Zavitkovski 1977a). The source was the same 10,000 curie of Cs-137 that had been used in the El Verde project in the Puerto Rican rain forest. It had been transported to the Oak Ridge National Laboratory, where it was refurbished, and then moved to the Wisconsin study site. The research design followed AEC's early forest projects with studies on radiation patterns and dosimetry along with studies of plant and animal populations and forest productivity. The radiation source killed or severely damaged radiosensitive species and changed the microclimate in the high-radiation zone. Zonation around the radiation source (Table 10.1) was essentially a plot of radiosensitivity, but it also represented a dissection of the structure of the forest ecosystem as had been found in the irradiation of the temperate forest at Brookhaven, New York (see Chapter 8).

Overall, the project provided much useful information about forest ecology but one is left with a sense that accomplishments for the research were less than had been originally hoped for. The degree to which the project had been truncated by the AEC's "changing priorities" is perhaps reflected by comparing the length of the project reports (361 total pages) when compared to earlier AEC publications for El Verde (1644 pages) and the Chariot Project (1250 pages). As a further indication of administrative changes taking place during the project, the pre-irradiation studies (Rudolph 1974) were published by the AEC while the post-irradiation studies (Zavitkovski 1977a) were published by ERDA, which had been split off from the AEC while the project was taking place.

TABLE 10.1 Zonation patterns formed in the northern forest in Rhinelander, Wisconsin, by a 10,000 curie ^{137}Cs gamma radiation source (Zavitkovski 1977b).

Exposure, r/20 Hour Day	*Distance From the Source, Meters*	*Vegetation Zone*
>1500	<5	lichens, some rushes (*Juncus* and *Luzula*), herbs
500–1500	5–10	sedge (*Carex*), *Lycopodium*, some *Salix* (willow)
250–500	10–14	herbs and *Rubus*
150–250	14–18	*Vaccinium*, *Acer saccharum*, *Carex*, *Aster*
65–150	18–25	understory shrubs and herbaceous layer
20–65	25–40	original forest without conifer trees
<20	>40	original forest with conifer trees

The Chernobyl Accident

With the changes to the AEC in the early 1970s Radioecology, as a discipline in the U.S., settled into a kind of quiescent period of monitoring for radioactive and thermal pollution. No more large-scale ecosystem projects or assessments were funded by the AEC's successor, ERDA. Smaller-scale Radioecology stayed active at some of the national labs, such as Oak Ridge and Savannah River, and at some academic programs, such as Colorado State University, but the days of peak funding by the AEC for ecosystem science were over. ERDA did support a number of focused ecology conferences in the 1970s which were showcases for the current U.S. Radioecology research. Emphasis had definitely shifted from research on impacts caused by nuclear explosives to impacts associated with commercial nuclear reactors. This state of the art in Radioecology lasted more than a decade, but changed dramatically on April 26, 1986, with the nuclear reactor accident at Chernobyl, a small city in Ukraine. As described by Bergeron (2002), "The accident was unbelievably violent, spewing dangerous radioactivity into the atmosphere, where it spread all over the world." There had been accidents around the world at nuclear reactors before this time (Webb 1976, Helgerson 1988), but they had not caused the kind of significant environmental damage, or the human mortality, that occurred at Chernobyl. Examples of reactor accidents in the U.S. were the Fermi plant near Detroit, Michigan, in 1964 (Fuller 1975), and the Three Mile Island plant near Harrisburg, Pennsylvania, in 1979 (Walker 2004).

Operator error and design flaws caused a reactor meltdown at Chernobyl. A large loss of coolant in the reactor's core led to an explosion that blew the roof off the containment building. The explosion sent debris all over the power plant site. Fire from the explosion burned out of control and the uranium fuel in the core melted and released a radioactive cloud of smoke into the sky. Plokhy (2018) noted that "altogether 50 million curies of radiation were released by the Chernobyl explosion, the equivalent of 500 Hiroshima bombs".

Over time the radioactivity in the atmosphere spread around the Northern Hemisphere. The highest fallout was over Russia, eastern Europe and southern Scandinavia, but the distribution was uneven due to complicated wind patterns. There was definitely a global impact from the accident at Chernobyl (Anspaugh et al. 1988): fallout was measured in a number of sites far from the Ukraine, such as Greenland (Davidson et al. 1987), and even the Chesapeake Bay watershed in the U.S. Mid-Atlantic region (Dibb and Rice 1988). Locally, radioactive fallout was not uniformly deposited around the reactor but it was carried in concentrated plumes called the Northern and Western Traces (see Chesser and Baker 2006 for a detailed map). The fallout included both highly radioactive fuel particles termed "hot particles" and lower-activity condensation particles, more similar to those found in global fallout after nuclear weapons testing (Beresford et al. 2016a). In general a mosaic of radioactive habitats and unaffected areas was created by the local fallout in the vicinity of the reactor. I-131, Cs-137, Sr-90 and Pu-239 were common radioactive nuclides in the fallout from the Chernobyl reactor.

Within a few days after the accident, the local population was evacuated and a fenced exclusion zone was set up. This zone was originally a circular area with a 10 km radius around the accident site. Later, once accurate radiation measurements were made, the exclusion area was expanded to include more area in the Ukraine and in Belarus. The final version of the exclusion zone was irregularly shaped and about the size of the U.S. state of South Carolina. This exclusion zone became an ecological reserve which the Soviets intended as "a unique laboratory which makes it possible to study effects on radiation on living nature" (D.D. 1988). The fires at the accident site were eventually extinguished and a concrete covering was constructed over the nuclear reactor, the first version of which was termed a "sarcophagus".

Much research at Chernobyl has been carried out on human exposure and contamination of the food production systems. Estimates of the death toll of local people exposed to radiation are controversial and range widely, from less than 100 to many thousands, if the toll of future cancer cases are factored in. Research on remediation of both agriculture and natural vegetation began soon after the accident and is ongoing (Davydchuk 1997).

Soviet scientists began studying the local environmental damage caused by the accident immediately, but information on the site was kept classified for several years (Savchenko 1995). The greatest damage was to the pine forests (*Pinus sylvestris* and *Pinus excella*) west of the nuclear power plant, which trapped radioactive fallout from the initial explosion in their canopies. This area was called the "red forest" because the needles turned an auburn color within several days as the trees died (Medvedev 1994). Much of this forest was later cut and buried in trenches in an effort to reduce the radioactivity. Some of the most contaminated area has been colonized by less radiosensitive plant species that survived the radioactive pollution, such as birch (*Betula perdula*) and several other hardwood trees (Sokolov et al. 1993, LeComte-Paradines et al. 2020). Savchenko (1995) provided a classification for forest damage with the impact directly proportional to radiation exposure (Table 10.2).

Over time a massive amount has been written about the nuclear reactor accident at Chernobyl, including a variety of articles and books for both popular audiences and the scientific community. Ecological research has been conducted using fieldwork and simulation models by scientists from a number of countries. As an example of the literature, to mark the 30th anniversary of the accident, 30 key papers were selected from a single journal and annotated (Beresford et al. 2016b). For terrestrial ecosystems one can find many published studies on microbes, lower plants (lichens and mosses), fungi, soil invertebrates and a diversity of vertebrate species, including wolves and bears. Aquatic ecosystems in the vicinity of Chernobyl have also been studied but to a lesser degree, perhaps because they occupy a much smaller area than terrestrial systems in the contaminated zone and because the exposure doses were lower for aquatic versus terrestrial organisms (Fesenko et al. 2005).

At the ecosystem scale, distributions and movements of radionuclides, especially Cs-137, have been monitored for forests both around the accident site and in other locations affected by long-distance fallout (Linkov and Schell 1999, Warner and Harrison 1993). Also, several computer models of Cs-137 dynamics in forests were developed and simulated for predictions of radioactivity levels into the future (Avila et al. 2001, Schell et al. 1996). There has been a special emphasis on genetics research with the goal of establishing damage caused by exposure to radiation. For example, Møller and Mousseau (2006) reviewed more than 25 studies from Chernobyl in cytogenetics, genetic variability and mutations. Geras'kin and his colleagues' (2013, 2014, 2016) research on Scot pine (*Pinus sylvestris*) illustrates this work and documents cytogenetic damage from exposure that may inhibit reproduction in plants from the most contaminated

TABLE 10.2 Zonation classification of impact of radiation exposure on vegetation around the Chernobyl nuclear reactor accident (from Savchenko 1995). Estimates of absorbed dose are from October 1, 1986.

Zone	*Nature of Impact*	*Absorbed Dose*
Zone of lethal impact	complete mortality of pine trees, different degrees of damage to leaf-bearing trees	80–100 Gy
Zone of sublethal impact	mortality of growth tips of larger trees, young trees died, abnormal reproductive organs, foliage turned yellow	10–20 Gy
Zone of average impact	tree growth process seriously suppressed, needle drop on pine trees	4–5 Gy
Zone of low impact	change in the formation of the generative organs, growth process suppressed	1–1.5 Gy

sites. Because some organisms have shown resistance to radiation, experimental research has been carried out with plants (Kovalchuk et al. 2004) and mice (Rodgers and Holmes 2008) that were cultured in the contaminated zone in order to understand this adaptation.

The most contaminated areas around the reactor had extremely high dose rates initially. Yet there is a lack of consensus on the impacts of longer-term chronic dose radiation (Beresford et al. 2016a, Lecomte-Pradines et al. 2020). To some extent this is because of the internal and external doses are difficult to determine or even to estimate (Chesser and Baker 2006). In the decades since the accident, there are some clear examples of species that have been impaired by the radiation from the accident (Møller et al. 2007), but there are also many examples of species that seem to be thriving in the exclusion zone (Webster et al. 2016, Deryabina et al. 2015, Baker et al. 1996). Controversial results have even been found in studies of the leaf litter decomposition rates in the forest of the exclusion zone (Bonzom et al. 2016, Mousseau et al. 2014). Clearly more research is needed to fully understand the genetic and ecological impacts of the Chernobyl nuclear reactor accident.

Even though there is some degree of controversy in the scientific literature about impacts, it is obvious that the ecosystems of the exclusion zone have come back after the disaster, essentially through self-organization. Photographs taken over a 20-year period by McMillan (2018) show the regrowth of vegetation at the accident site and provide visual documentation of the restoration of the ecosystem. This regrowth is amazing, given the amount of devastation that occurred when the nuclear reactor exploded and contaminated a huge landscape. There was some effort to restore the ecosystem, but most of the regrowth was by self-organization with seeding from patches in the contaminated zone that were shielded from radiation. This is a remarkable example of the ability of the ecosystem to restore itself after a major disturbance. There may not be as much "ecological integrity" as before the accident, but the ecosystems have shown resiliency to the accident. Baker and Chesser (2000) describe the exclusion zone as "a vibrant ecosystem" and "beautiful" and they describe "numerous sightings of moose (*Alces alces*), roe deer (*Capreol capreolus*), Russian wild boar (*Sus scrofa*), foxes (*Vulpes vulpes*), river otter (*Lutra canadensis*), and rabbits (*Lepus europaeus*) with the 10-km exclusion zone".

The removal of humans from the exclusion zone is a contributing factor in the regrowth of the ecosystem, as noted by Beresford et al. (2016a), who suggest there are "major continuing ecosystem changes brought about in the exclusion zone due to the removal of the human population and the cessation of associated agricultural, forestry, hunting and fishing activities". The term "anthropause" has been coined to describe this effect, based on experience during the recent lockdown associated with the Covid-19 pandemic (Weathers et al. 2021), but the phenomenon was noted in the exclusion zone around Chernobyl as early as a decade after the accident (Williams 1995). Figure 10.1 shows the stages of ecosystem development that occurred at Chernobyl with

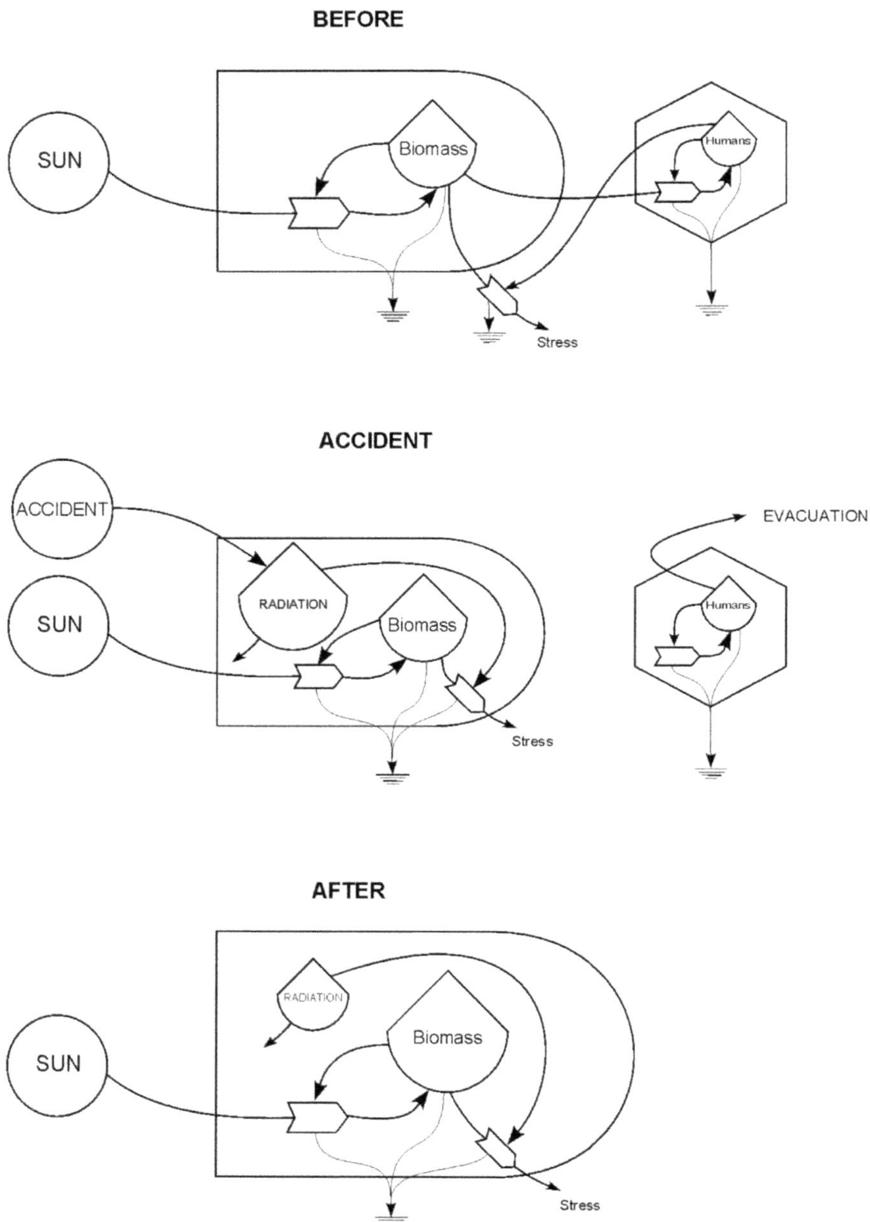

FIGURE 10.1 Overview models of the consequences of the Chernobyl nuclear accident, illustrating the anthropause concept when humans are disconnected from the ecosystem.

minimodels in the Energy Circuit Language (Odum 1994), illustrating the anthropause effect. Before the accident, as shown at the top of the figure, humans in the local area drained biomass from the ecosystem with stress, such as pollution, and with direct consumption from activities, such as harvest and hunting. The accident, as shown in the middle of the figure, added a large amount of radiation to the ecosystem, which drained biomass due to stress, but the local humans were evacuated and so they no longer interacted with the ecosystem. Finally after decades, as shown in the bottom of the figure, the ecosystem has restored itself with greater biomass than before the accident since the radiation stress has diminished due to radioactive decay and there is no negative impact from humans since they are still removed from the local area. At least in a qualitative sense, the significance of this example of ecosystem resiliency at Chernobyl is that life and the ecosystem can survive even a terrible nuclear accident.

The Fukushima Accident

Twenty-five years after the accident at Chernobyl another major nuclear reactor accident occurred, on the east coast of Japan, north of Tokyo in the Fukushima Prefecture. A magnitude 9.0 earthquake and a 40-meter-high tsunami on March 11, 2011, triggered meltdowns at the Fukushima Daiichi Nuclear Power Plant. Flooding resulted in a complete loss of power at the plant. Without power the operators were not able to cool the reactors and, thus, a core meltdown took place. As a consequence an explosion occurred, which ruptured the containment structure, releasing a cloud of radioisotopes into the atmosphere. I-131, Cs-134 and Cs-137 were some of the most significant elements in the cloud, which was equivalent to about one-tenth of the radioactivity released in the Chernobyl accident (Steinhauser et al. 2014). About 70–80% of the vented radiation blew out over the adjacent Pacific Ocean, with the rest being deposited over the land (see Yoshida and Kanda 2012 for a map of the cesium fallout distribution in Japan). Radioactivity spread around the globe in the Northern Hemisphere as shown, for example, by the food chain contamination of milk by I-131 as far away as California (Treinen 2015) and France (Parache et al. 2011). Approximately 150,000 people were evacuated from around the power plant and an exclusion zone was set up. Although there were no immediate radiation-related deaths from the accident, many deaths have been attributed to the evacuation process, especially among the elderly. Furthermore, elevated rates of thyroid cancer suggest that the mortality estimates will increase (Normile 2016). The power at the site was restored within a week after the accident and the damaged reactors were stabilized. However, leakage of water contaminated with radioactivity into the Pacific continued for multiple years. An early overview of the accident is given by Povinec et al. (2013).

A diverse program of Radioecology research began immediately after the Fukushima accident, carried out mostly by Japanese scientists along with a few foreigners, including some with past experience at Chernobyl (see for an overview, Hashimoto et al. 2022). A number of studies were done on radiation exposures of individual plant and animal species and examples of genetic damage related to exposure were documented (see, for example, Kubota et al. 2015, Taira et al. 2014, Gombeau et al. 2020, Yoschenko et al. 2016). Forest ecosystem studies of inventories and movements of radiocesium from fallout were carried out and fast rate processes were documented (Komatsu et al. 2016). A review of connections between forests and streams was done by Sakai et al. (2021), illustrating the landscape scale of radiation impacts. Radioactivity released into the Pacific Ocean was carried by currents away from Japan and effectively diluted (Buesseler et al. 2011, 2012). However, migratory fish with elevated radiocesium content were found as far away as California (Madigan et al. 2013) and potential biomagnification of the contaminants through ocean food webs was a concern (Alava and Gobas 2016).

The Fukushima nuclear accident occurred only about 10 years ago at the time this book was written, so long-term trends in the response of the contaminated terrestrial ecosystems of the exclusion area have yet to emerge. However, the ecosystems in the exclusion seem to be intact and functioning (Lyons et al. 2020). These ecosystems present opportunities for study of the continuing ecological impacts of a nuclear accident in a similar fashion to what has been done in the exclusion zone around Chernobyl.

Comparisons of the "Old" and "New" Radioecology

The nuclear accidents at Chernobyl and Fukushima definitely rejuvenated the field of Radioecology, as noted by Beresford et al. (2016b):

> The Chernobyl accident led to a large resurgence in radioecological studies both to aid remediation and to be able to make future predictions on the post-accident situation, but, also, in recognition that more knowledge was required to cope with future accidents.

However, although the field has clearly been revitalized, the nature of the research has changed from what the AEC funded during the early years of the Cold War, which is the main subject of this book. The AEC-funded Radioecology research focused on the effects of nuclear bomb detonations that would occur during war or in peaceful uses such as in nuclear excavations envisioned in Project Plowshare. Commercial nuclear reactors for generating electricity had not been designed or constructed in the early years of the AEC so Radioecology applied to nuclear accidents came much later. Moreover, the impacts of a reactor meltdown were hardly imaginable before they actually occurred in the accident at Chernobyl in 1986. Most of the AEC-supported Radioecology research involved controlled studies of nuclear detonations, such as at the PPG or the NTS, or focused field studies where a gamma radiation source or radionuclides were experimentally introduced into natural ecosystems. These were basically small-scale projects that typically played out over only a few years. In contrast, the Radioecology research in regard to nuclear accidents has been uncontrolled, in the context of experimental design, with effects occurring over large areas and long time periods. The AEC-funded research was not done in response to a massive, acute emergency, like a reactor meltdown, but rather it was the consequence of a carefully crafted plan which was carried out over nearly 20 years at a number of field sites. Because of these fundamental differences, there is relatively little that can be compared between the "old" Radioecology funded by the AEC in the early years of the Cold War and the "new" Radioecology of Chernobyl and Fukushima during the post–Cold War era.

Because of the differences, it is perhaps not surprising that the old Radioecology did little to inform the new Radioecology involving nuclear accidents. Evidence of this disconnect can be seen by the lack of citation of the old AEC-supported Radioecology research in the publications about the nuclear accidents. A few recent papers about Chernobyl and Fukushima cite the work done at Oak Ridge on the Cesium Forest in the 1960s but the other projects on temperate forests, such as at Brookhaven or Wisconsin, do not appear in the literature cited of the recent publications. One exception is Geras'kin's (2016) review of the effects of ionizing radiation on ecosystems, which does include a summary table with several of the old AEC-supported Radioecology projects. To some extent the disconnect may have occurred because some of the important concepts of ecosystem structure and function, in particular as they relate to mineral cycling, that were developed and refined during the AEC-funded projects of the 1950s and 1960s have become common knowledge within Ecology to the degree that it is no longer necessary to cite them in a literature review of a recent publication.

One of the biggest differences between the old and new Radioecologies is the funding mechanisms. Even though the accidents at Chernobyl and Fukushima were major environmental disasters, it was difficult for researchers to obtain funding for even basic studies about them. As an example of this problem, Table 10.3 lists funding sources acknowledged in 15 papers published in scientific journals about Radioecology at Chernobyl that were chosen more or less at random. Twenty-five funding agencies representing a wide variety of countries are listed as providing support for the research. A similar table done for U.S. Radioecology research done in the 1950s and 1960s would have only the Atomic Energy Commission as the funding source. This difference in funding between the old and new Radioecologies has contributed to a fragmentation and a lack of synthesis in the recent research about the nuclear reactor accidents compared to the older research carried out in the U.S. during the early years of the Cold War. Some researchers have complained about the difficulty in attracting funds for work on nuclear reactor accidents (Møller and Mousseau 2006), even to the extent that

TABLE 10.3 Funding sources listed in the acknowledgments section from a sample of 15 research papers selected haphazardly from the literature on Chernobyl.

Funding Source	*Frequency of Occurrence*
European Commission	4
SSI contract (Russian)	1
U.S. Nuclear Regulatory Commission	2
U.S. National Academy of Sciences under CAST program	1
Royal Society International Exchange Fund	1
UK Natural Environment Research Council	6
UK Radioactive Waste Management Ltd.	5
Environment Agency of England and Wales	1
Japan Atomic Energy Agency	1
INTAS-2001–0556 "RESPOND" (Russian)	1
U.S. Department of Energy	5
CNRS (France)	1
University of South Carolina School of the Environment	1
Bill Murray and the Samuel Freeman Charitable Trust	1
U.S. Civilian Research and Development Foundation	1
National Science Foundation (U.S.)	1
National Geographic Society	3
Ministry of Ecology and Natural Resources of Ukraine	1
EURATOM COMET project	1
Institut de Radioprotection et de Surete Nucleaire	1
Norwegian Radiation Protection Authority	1
The Royal Society	1
Russian Academy of Sciences	1
Russian Foundation for Basic Research	1
Norwegian Research Council	1

Voigt (2001) wondered if Radioecology was a "dying field" due to the relative lack of support.

Nuclear Winter

As a final aside, a story related to Radioecology evolved in the 1980s around possible climate change that would be caused by large-scale nuclear war. The thinking was that dust and smoke from fires caused by nuclear detonations would fill the atmosphere, thereby cutting off sunlight for significant amounts of time. The decline in sunlight would cause temperatures to drop and would reduce photosynthesis in natural and agroecosystems. It was called "nuclear winter" due to the temperature drop and the reductions in plant production, as would typically occur during the normal winter season. This kind of indirect impact from nuclear war had been overlooked in previous assessment that focused on direct effects. The theory came from an unusual source—the astronomer Carl

Sagan—who put together a team of scientists who modeled atmospheric impacts of nuclear war (Turco 1997). As noted by Sagan (1984),

> We stumbled upon these results by accident, by a circuitous route, by one of those circumstances common in science where studying something purely for its intellectual interest leads you to conclusions of surprising practical utility. For me, it began in 1971 with the Mariner 9 exploration of the planet Mars. Mariner 9 was the first spacecraft to orbit another planet. Its engineers guaranteed that it would work only for three months after orbital injection. The spacecraft arrived at Mars to find the planet completely covered with a global dust storm. After a month of photographing an almost entirely featureless disk, we began to worry seriously that by the time the dust would settle out of the Martian atmosphere the spacecraft would no longer be working. The dust storm in fact took three months to dissipate, but the spacecraft worked far better than the engineers had said—and for the next year we were able to examine the planet pole to pole in the first detailed orbital reconnaissance of another planet. . . . We were able to see the temperature of the atmosphere and that of the surface change with time. The results showed that the atmosphere was considerably warmer than is usually the case for Mars, and the surface considerably colder. As the dust settled out, the atmosphere became cooler and the surface warmer—both approaching their usual, or "ambient" values. It was not difficult to understand the reasons for this. The winds had stirred a great deal of dust off the Martian deserts into the atmosphere. Sunlight was being absorbed by the high-altitude dust, thereby heating the atmosphere. But, by the same token, the sunlight was impeded from reaching the surface, and so the surface had cooled. . . . There are differences between Mars and Earth, of course—including the absence of oceans on Mars, and the fact that it is farther away from the sun. But it seemed to us that the Martian experience might be relevant to Earth."

A decade later Sagan was reminded of the Mariner 9 experience, when reading two new studies. First was Alvarez et al.'s (1980) theory of the demise of dinosaurs 65 million years ago caused by an asteroid that struck the Earth, raising a cloud of debris into the atmosphere which blocked sunlight from the surface for perhaps a year; and the second was the study by Crutzen and Birks (1982) which estimated the amount of smoke that could be injected into the atmosphere by a large-scale nuclear war. Sagan and his team synthesized all of these studies into the nuclear winter theory (Turco et al. 1983). The impacts from nuclear winter were severe, at the extreme perhaps causing the extinction of the human species. Because of the new projections on the magnitude of direct and indirect impacts of large-scale nuclear war and because of Sagan's notable reputation and skills as a communicator, nuclear winter theory attracted great attention from the general public, from scientists and even from government officials in the U.S.

and in the Soviet Union. Conferences were held and books were published on nuclear winter in short order. Projections by Sagan's group were criticized by modelers from the U.S. National Center for Atmospheric Research and others who suggested the impacts would be less severe, perhaps better termed a "nuclear autumn". Debate occurred (see, for example, Marshall 1987) and new models were constructed with refinements on the original theory (Sagan and Turco 1990). Overall, nuclear winter studies lasted only about a decade, and interest in the subject waned with the conclusion of the Cold War by about 1991.

Research from the AEC-funded ecosystem studies from the 1950s and 1960s contributed little to the nuclear winter theory. George Woodwell was the only scientist from the AEC days to contribute, but he was an active participant (Woodwell 1983, 1985). To some extent the lack of overlap is not surprising because the 1950s and 1960s AEC ecological research consisted of detailed monitoring of single nuclear detonations during tests or carefully designed experimental radiation studies, while the nuclear winter research focused on computer simulations of what would happen after a nuclear war with hundreds or even thousands of detonations taking place at essentially the global scale. On the one hand, nuclear winter researchers were often motivated by humanitarian concerns to save humanity from destruction and were actively trying to stop nuclear proliferation. For example, Carl Sagan was arrested for protesting against nuclear testing at the Nevada Test Site (Poundstone 1999). On the other hand, scientists funded by the AEC in ecosystem research were generally less politically motivated.

Literature Cited

Alava, J. J. and F. A. P. C. Gobas. 2016. Modeling ^{137}Cs bioaccumulation in the salmon-resident killer whale food web of the Northeastern Pacific following the Fukushima nuclear accident. Science of the Total Environment 544:56–67.

Alvarez, L. W., W. Alvarez, F. Asaro and H. V. Michel. 1980. Extraterrestrial cause for the Cretaceous-Tertiary extinction. Science 208:1095–1108.

Anspaugh, L. R., R. J. Catlin and M. Goldman. 1988. The global impact of the Chernobyl reactor accident. Science 242:1513–1519.

Auerbach, S. I. 1971. Ecological considerations in siting nuclear power plants: The long-term biotic effects problem. Nuclear Safety 12:25–34.

Auerbach, S. I., R. W. Brocksen, R. B. Craig, F. O. Hoffman, S. V. Kaye, D. E. Reichle and E. G. Struxness. 1976. Environmental impact statements. Science 193:188–189.

Avila, R., R. Bergman, M. Scimone, S. Fesenko, N. Sancharova and L. Moberg. 2001. A comparison of three models of ^{137}Cs transfer in forest ecosystems. Journal of Environmental Radioactivity 55:315–327.

Baker, R. J. and 10 co-authors. 1996. Small mammals from the most radioactive sites near the chornobyl nuclear power plant. Journal of Mammalogy 77:155–170.

Baker, R. J. and R. K. Chesser. 2000. The Chornobyl nuclear disaster and subsequent creation of a wildlife preserve. Environmental Toxicology and Chemistry 19:1231–1232.

Beatley, J. C. 1976. Vascular Plants of the Nevada Test Site and Central-Southern Nevada: Ecological and Geographic Distributions. Technical Information Center, Energy Research and Development Administration, Washington, DC.

Beresford, N. A., S. Fesenko, A. Konoplev, L. Skuterud, J. T. Smith and G. Voigt. 2016a. Thirty years after the Chernobyl accident: What lessons have we learnt? Journal of Environmental Radioactivity 157:77–89.

Beresford, N. A., S. Fesenko, A. Konoplev, J. T. Smith, L. Skuterud and G. Voigt. 2016b. Thirty years after the Chernobyl accident: 30 key papers published in the Journal of Environmental Radioactivity. Journal of Environmental Radioactivity 157:38–40.

Bergeron, K. D. 2002. Tritium on Ice: The Dangerous New Alliance of Nuclear Weapons and Nuclear Power. MIT Press, Cambridge, MA.

Bonzom, J.-M. and 10 co-authors. 2016. Effects of radionuclide contamination on leaf litter decomposition in the Chernobyl exclusion zone. Science of the Total Environment 562:596–603.

Buesseler, K., M. Aoyama and M. Fukasawa. 2011. Impacts of the Fukushima Nuclear Power Plants on marine productivity. Environmental Science & Technology 45:9931–9935.

Buesseler, K. O. and 13 co-authors. 2012. Fukushima-derived radionuclides in the ocean and biota off Japan. Proceedings of the National Academy of Sciences (PNAS) 109:5984–5988.

Chesser, R. K. and R. J. Baker. 2006. Growing up with Chernobyl. American Scientist 94:542–549.

Crutzen, P. J. and J. W. Birks. 1982. The atmosphere after a nuclear war: Twilight at noon. Ambio 11: 114–125.

Davidson, C. I., J. R. Harrington, M. J. Stephenson et al. 1987. Radioactive cesium from the Chernobyl accident in the Greenland Ice Sheet. Science 237:633–634.

Davydchuk, V. 1997. Ecosystem remediation in radioactively polluted areas: The Chernobyl experience. Ecological Engineering 8:325–336.

D.D. 1988. Chernobyl area to be ecological reserve. Science 240:877.

Deryabina, T. G., S. V. Kuchmel, L. L. Nagorskaya, T. G. Hinton, J. C. Beasley, A. Lerebours and J. T. Smith. 2015. Long-term census data reveal abundant wildlife populations at Chernobyl. Current Biology 25:R811–R826.

Dibb, J. E. and D. L. Rice. 1988. Chernobyl fallout in the Chesapeake Bay region. Journal of Environmental Radioactivity 7:193–196.

Eichholz, G. G. 1976. Environmental Aspects of Nuclear Power. Ann Arbor Science Publ., Ann Arbor, MI.

Fesenko, S. V., R. M. Alexakhin, S. A. Geras'kin, N. I. Sanzharova, Ye. V. Spirin, S. I. Spiridonov, I. A. Gontarenko and P. Strand. 2005. Comparative radiation impact on biota and man in the area affected by the accident at the Chernobyl Nuclear Power Plant. Journal of Environmental Radioactivity 80:1–25.

Fuller, J. G. 1975. We Almost Lost Detroit. Ballantine Books, New York, NY.

Geras'kin, S. A. 2016. Ecological effects of exposure to enhanced levels of ionizing radiation. Journal of Environmental Radioactivity 162–163:347–357.

Geras'kin, S. A., N. S. Dikareva, A. A. Oudalova, D. V. Vasil'ev and P. Yu. Volkova. 2016. The consequences of chronic radiation exposure of Scots Pine in the remote period after the Chernobyl accident. Russian Journal of Ecology 47:26–38.

Geras'kin, S. A., T. Evseeva and A. Oudalova. 2013. Effects of long-term chronic exposure to radionuclides in plant populations. Journal of Environmental Radioactivity 121:22–32.

Geras'kin, S. A. and P. Yu. Volkova. 2014. Genetic diversity in Scots pine populations along a radiation exposure gradient. Science of the Total Environment 496:317–327.

Gillette, R. 1971. AEC's new environmental rules for nuclear plants may open new debate, extend delays, raise plant costs. Science 173:1112–1113.

Glasstone, S. and W. H. Jordan. 1980. Nuclear Power and Its Environmental Effects. American Nuclear Society, La Grange Park, IL.
Gombeau, K. and eleven co-authors. 2020. Dose-dependent genomic DNA hypermethylation and mitochondrial DNA damage in Japanese tree frogs sampled in the Fukushima Daiichi area. Journal of Environmental Radioactivity 225:106429.
Hashimoto, S., M. Komatsu and S. Miura (eds.). 2022. Forest Radioecology in Fukushima, Radiocesium Dynamics, Impact, and Future. Springer, Singapore.
Helgerson, J. 1988. Nuclear Accidents. Franklin Watts Publ., New York, NY.
Komatsu, M. and 12 co-authors. 2016. Characteristics of initial deposition and behavior on radiocesium in forest ecosystems of different locations and species affected by the Fukushima Daiichi Nuclear Power Plant accident. Journal of Environmental Radioactivity 161:2–10.
Kovalchuk, I., V. Abramov, I. Pogribny and O. Kovalchuk. 2004. Molecular aspects of plant adaptation to life in the Chernobyl zone. Plant Physiology 135:357–363.
Kubota, Y. and 17 co-authors. 2015. Chromosomal aberrations in wild mice captured in areas differentially contaminated by the Fukushima Dai-Ichi Nuclear Power Plant accident. Environmental Science & Technology 49:10074–10083.
Lecomte-Pradines, C., C. Adam-Guillermin, S. Gashchak, C. Bradshaw, D. Copplestone, N. A. Beresford. 2020. More than thirty years after the Chernobyl accident: What do we know about the effects of radiation on the environment? Journal of Environmental Radioactivity 211:106108.
Linkov, I. and W.R. Schell (eds.). 1999. Contaminated Forests, Recent Developments in Risk Identification and Future Perspectives. Kluwer Academic Publishers, Dordrecht.
Lyons, P. C., K. Okuda, M. T. Hamilton, T. G. Hinton and J. C. Beasley. 2020. Rewilding of Fukushima's human evacuation zone. Frontiers in Ecology and the Environment 18:127–134.
Madigan, D. J., Z. Baumann, O. E. Snodgrass, H. A. Ergül, H. Dewar and N. S. Fisher. 2013. Radiocesium in Pacific bluefin tuna *Thunnus orientalis* in 2012 validates new tracer technique. Environmental Science & Technology 47:2287–2294.
Marshall, E. 1987. Nuclear winter debate heats up. Science 235:271–273.
McMillan, D. 2018. Growth and Decay, Prypiat and the Chernobyl Exclusion Zone. Steidl Publ., Göttingen, Germany.
Medvedev, Z. A. 1994. Chernobyl: Eight years after. Trends in Ecology and Evolution 9:369–371.
Møller, A. P. and T. A. Mousseau. 2006. Biological consequences of Chernobyl: 20 years on. Trends in Ecology and Evolution 21:200–207.
Møller, A. P., T. A. Mousseau, F. de Lope and N. Saino. 2007. Elevated frequency of abnormalities in barn swallows from Chernobyl. Biology Letters 3:414–417.
Mousseau, T. A., G. Milinevsky, J. Kenney-Hunt and A. P. Møller. 2014. Highly reduced mass loss rates and increased litter layer in radioactively contaminated areas. Oecologia 175:429–437.
Normile, D. 2016. Epidemic of fear. Science 351:1022–1023.
Odum, H. T. 1994. Ecological and General Systems: An Introduction to Systems Ecology. University Press of Colorado, Niwot, CO.
Parache, V., L. Pourcelot, S. Roussel-Debet, D. Orjollet, F. Leblanc, C. Soria, R. Gurriaran, Ph. Renaud and O. Masson. 2011. Transfer of ^{131}I from Fukushima to the vegetation and milk of France. Environmental Science & Technology 45:9998–10003.
Plokhy, S. 2018. Chernobyl: The History of a Nuclear Disaster. Basic Books, New York, NY.

Poundstone, W. 1999. Carl Sagan: A Life in the Cosmos. Henry Holt and Co., New York, NY.

Povinec, P. P., K. Hirose and M. Aoyama. 2013. Fukushima Accident, Radioactivity Impact on the Environment. Elsevier, Amsterdam.

Rodgers, B. E. and K. M. Holmes. 2008. Radio-adaptive response to environmental exposures at Chernobyl. Dose-Response 6:209–221.

Rolph, E. S. 1979. Nuclear Power and the Public Safety. Lexington Books, Lexington, MA.

Rudolph, T. D. (ed.). 1974. The Enterprise, Wisconsin, Radiation Forest: Preirradiation Ecological Studies. TID-26113, Technical Information Center, U. S Atomic Energy Commission, Washington, DC.

Sagan, C. 1984. The atmospheric and climatic consequences of nuclear war. Pp. 3–39. In: P. R. Ehrlich, C. Sagan, D. Kennedy and W. O. Roberts. The Cold and the Dark, The World After Nuclear War. W. W. Norton & Co., New York, NY.

Sagan, C. and R. Turco. 1990. A Path Where No Man Thought, Nuclear Winter and the End of the Arms Race. Random House, New York, NY.

Sakai, M., H. Tsuji, Y. Ishii, H. Ozaki, S. Takechi, J. Jo, M. Tamaoki, S. Hayashi and T. Gomi. 2021. Untangling radiocesium dynamics of forest-stream ecosystems: A review of Fukushima studies in the decade after the accident. Environmental Pollution 288:117744.

Savchenko, V. K. 1995. The Ecology of the Chernobyl Catastrophe. Man in the Biosphere Series No. 16, Parthenon Publ., Lanes, UK.

Schell, W. R., I. Linkov, V. Rimkevich, O. Chistic, A. Lutsko, A. M. Dvornik and T. A. Zhuchenko. 1996. Model-directed sampling in Chernobyl forests: General methodology and 1994 sampling program. Science of the Total Environment 1996:229–240.

Schlichting, P. E., C. N. Love, S. C. Webster and J. C. Beasley. 2019. Efficiency and composition of vertebrate scavengers at the land-water interface in the Chernobyl Exclusion Zone. Food Webs 18:e00107.

Seymour, A. H. 1976. Dedication of symposium to professor John N. Wolfe. Pp. xiii–xv. In: C. E. Cushing, Jr. (ed.). Radioecology and Energy Resources, Proceedings of the Fourth National Conference. Dowden, Hutchinson & Ross, Inc., Stroudsburg, PA.

Sokolov, V. E., I. N. Rjabov, I. A. Ryabtsev, F. A. Tikhomirov, V. A. Shevchenko and A. I. Taskaev. 1993. Ecological and genetic consequences of the Chernobyl atomic power plant accident. Vegetatio 109:91–99.

Sprugel, G., Jr. 1975. John N. Wolfe, 1910–1974: Resolution of respect. Bulletin of the Ecological Society of America 56:16–22.

Steinhauser, G., A. Brandl and T. E. Johnson. 2014. Comparison of the chernobyl and fukushima nuclear accidents: A review of the environmental impacts. Science of the Total Environment 470–471:800–817.

Taira, W., C. Nohara, A. Hiyama and J. M. Otaki. 2014. Fukushima's biological impacts: The case of the pale grass blue butterfly. Journal of Heredity 105:710–722.

Transeau, E. N. 1926. The accumulation of energy by plants. Ohio Journal of Science 26:1–10.

Treinen, R. M. 2015. An analysis of the intake of Iodine-131 by a dairy herd post-Fukushima and the subsequent excretion in milk. Journal of Environmental Radioactivity 149:135–143.

Turco, R. P. 1997. Carl Sagan and nuclear winter. Pp. 239–246. In: Y. Terzian and E. Bilson (eds.). Carl Sagan's Universe. Cambridge University Press, Cambridge, UK.

Turco, R. P., O. B. Toon, T. P. Ackerman, J. B. Pollack and C. Sagan. 1983. Nuclear winter: Global consequences of multiple nuclear explosions. Science 222:1283–1292.
Voigt, G. 2001. Are radioecologists the dinosaurs of this century? Journal of Environmental Radioactivity 53:3–4.
Walker, J. S. 2004. Three Mile Island: A Nuclear Crisis in Historical Perspective. University of California Press, Berkeley, CA.
Warner, F. and R. M. Harrison (eds.). 1993. Radioecology after Chernobyl, Biogeochemical Pathways of Artificial Radionuclides. John Wiley & Sons, Chichester.
Weathers, K. C., D. Ojima, S. K. Collinge and O. Sala. 2021. Leveraging the anthropause. Frontiers in Ecology and the Environment 19:315.
Webb, R. E. 1976. The Accident Hazards of Nuclear Power Plants. University of Massachusetts Press, Amherst, MA.
Webster, S. C., M. E. Byrne, S. L. Lance, C. N. Love, T. G. Hinton, D. Shamovich and J. C. Beasley. 2016. Where the wild things are: Influence of radiation on the distribution of four mammalian species within the Chernobyl Exclusion Zone. Frontiers in Ecology and Environment 14:185–190.
Williams, N. 1995. Chernobyl: Life abounds without people. Science 269:304.
Wolfe, J. N. 1963. Impact of atomic energy on the environment and environmental science. Pp. 1–2. In: V. Schultz and A. W. Klement, Jr. (eds.). Radioecology, Proceedings of the First National Conference. Reinhold, New York, NY.
Wolfe, J. N. 1969. Radioecology: Retrospection and future. Pp. xi–xii. In: D. J. Nelson and F. C. Evans (eds.). Symposium on Radioecology, Proceedings of the Second National Symposium. CONF-670503, U.S. Atomic Energy Commission, Washington, DC.
Woodwell, G. M. 1983. The biotic effects of ionizing radiation. Pp. 131–139. In: J. Peterson (ed.). The Aftermath: The Human and Ecological Consequences of Nuclear War. Pantheon Books, New York, NY.
Woodwell, G. M. 1985. Energy enough to cleave the Earth. Bioscience 35:584–585.
Yoschenko, V., K. Nanba, S. Yoshida, Y. Watanabe, T. Takase, N. Sato and K. Kejtoku. 2016. Morphological abnormalities in Japanese red pine (*Pinus densiflora*) at the territories contaminated as a result of the accident at Fukushima Dai-Ichi Nuclear Power Plant. Journal of Environmental Radioactivity 165:60–67.
Yoshida, N. and J. Kanda. 2012. Tracking the Fukushima radionuclides. Science 336: 1115–1116.
Zavitkovski, J. (ed.). 1977a. The Enterprise, Wisconsin, Radiation Forest: Radioecological Studies. TID-26113-P2, Technical Information Center, Energy Research and Development Administration, Washington, DC.
Zavitkovski, J. 1977b. Effects of gamma radiation on a northern forest ecosystem: A summary. Pp. 197–206. In: J. Zavitkovski (ed.). The Enterprise, Wisconsin, Radiation Forest: Radioecological Studies. TID-26113-P2, Technical Information Center, Energy Research and Development Administration, Washington, DC.

11

SUMMARY AND CONCLUSIONS

Cold War–Era Radioecology and the Ecosystem Concept

Cold War–era Radioecology was a brief but significant pulse in the history of ecology as an academic discipline. Through the 1950s and 1960s the AEC funneled millions of dollars into support of ecological research to study the effects of detonation of nuclear weapons. This was necessary because radioactivity released in the detonations was a new, unknown form of pollution with potentially serious impacts to humans and all forms of biodiversity. At the start of the Cold War there was near complete uncertainty about the short- and long-term effects of detonating nuclear weapons. As an example of the magnitude of this uncertainty, before the test of the first bomb at the Trinity site in August 1945, one of the Manhattan Project scientists worried that there was a small probability that the detonation might catch the Earth's atmosphere on fire! Once atomic bombs started exploding at the test sites in the Pacific and in the Nevada desert it became evident that ecologists needed to work out what the ecosystem concept really meant. An understanding of the ecosystem concept was needed in order to assess the impacts of radiation and of fallout that were being released from the tests. The radioecologists took the concept written about by Tansley and Lindeman and turned it into something that could be quantified with field measurements and illustrated with systems diagrams. Ultimately, Radioecology studies funded by the AEC helped to reduce the uncertainty of environmental impacts from nuclear warfare, although many questions remain to this day.

There was an urgency for the research because the threat of nuclear war was imminent. While the AEC stumbled early to respond to the need for ecological research on radiation effects, especially before the tragic Bravo test at the Pacific Proving Grounds in 1954, it eventually became the dominant source of funding for Ecology for some two decades after World War II. Major Radioecology

DOI: 10.4324/9781003284901-14

projects were carried out in a coordinated effort in a variety of environments within U.S.-controlled territory, ranging from deserts to tropical rain forests and from the open ocean to small freshwater lakes and streams. Frank Golley (1993) suggested this was an "invisible college", in the sense of focused, shared purpose of research carried out by scientists at different locations. However, at the time there was little that was "invisible" about Radioecology research. There was a single funding source (the AEC) and to a large extent the researchers were encouraged to publish their results in the open literature. In this text, the research at seven major sites is reviewed, but there were many more projects included in AEC's Radioecology program. Just to mention a few of the others, early work on food chains (Davis and Foster 1958) and nutrient cycling (Whittaker 1961) was done at the Hanford site in eastern Washington State where most of the emphasis was on long-term monitoring of the site (Becker 1990) and contamination of the Columbia River estuary (Pruter and Alverson 1972). Robert Platt at Emory University, near Atlanta, Georgia, operated a large Radioecology facility (Platt 1965) with research on old fields (Daniel and Platt 1968) and granite outcrop communities (McCormick and Platt 1962), which tended to concentrate fallout radionuclides. He even invented a portable system for growing these unique communities on artificial surfaces to facilitate experimentation (Platt and McCormick 1964). Ward Whicker and others created a graduate program in Radioecology at Colorado State University in which they took advantage of both Rocky Mountain habitats to the west (Osburn 1967) and the short-grass prairie to the east (Fraley and Whicker 1971). The U.S. Fish and Wildlife Service operated its Radioecology Laboratory at Beaufort, North Carolina (Rice and Duke 1968) with studies in the mid-Atlantic estuaries, including a microcosm complex for research on radioactive zinc cycling in a salt marsh system (Duke et al. 1966). Tritium was added to a Lake Erie marsh in Ohio and its dynamics were monitored (Adams et al. 1976) and modeled (White et al. 1978). These and other programs contributed to the "invisible college" of Radioecology supported by the AEC.

A good overview sense of the depth and breadth of this research can be found in the first three National Conferences on Radioecology held in the U.S. (Schultz and Klement 1963, Nelson and Evans 1969, Nelson 1971). There was a great deal of exchange of personnel between the AEC research sites, which resulted in a sharing of ideas and techniques. For example, E. P. Odum led the work at the Savannah River Site, but he served as an advisor at Oak Ridge and visited the Pacific Proving Grounds for coral reef work and El Verde for collaboration with his brother, and the Nevada Test Site and Hanford during his fellowship; Frank McCormick got his M.S. degree working with Robert Platt at Emory University on Radioecology of granite outcrops, then worked at El Verde and Savannah River and later at the University of Tennessee and Oak Ridge; Peter Murphy got his MS degree working on the El Verde Project in the rain forest, then got his Ph.D. working on the granite outcrop microcosms with McCormick at the University of North Carolina, then worked in the Enterprise irradiation forest in Wisconsin after taking a faculty position at Michigan State University; but Clayton Gist probably holds the record by working at nearly every major Radioecology project over his career (see Chapter 4). These are just a few of the examples of exchanges that took place among AEC sites that resulted in a "cross-fertilization" of the ecosystem concept. There was also exchange with Soviet ecologists who were dealing with similar issues of environmental impacts of radiation in their nuclear program (Alexakhin 2006, Klechkovskii et al. 1973). The Soviets' concept of the ecosystem was slightly different, with less theory and abstraction, than that of the West (Johnson and French 1981), but they were strong in natural resources and biogeochemistry. The U.S. government sponsored translation of Soviet publications on Radioecology (Klement et al. 1968) and Soviet scientists occasionally participated in international conferences on the subject. Thus, information exchange did take place across the Iron Curtain despite ideological differences, though it probably had little influence compared to exchanges between the AEC-funded sites. The heritage of Cold War–era Radioecology is in the published research and in the human capital of many individual ecologists whose careers were boosted by the AEC research funding. The modern, post–World War II model of the ecosystem emerged from the different Radioecology projects and the personalities of the scientists who fashioned the model through their research.

In terms of the academic discipline of Ecology, the main contribution of Cold War–era Radioecology was the operationalization of the ecosystem concept. Most historians of the science of Ecology credit Arthur Tansley as the originator of the ecosystem concept, but he really just introduced it as an idea near the end of his career and he never elaborated on it. Raymond Lindeman took the ecosystem concept a giant step forward, right before World War II began, with his model of a small bog lake from his dissertation, but he died too young to be able to elaborate on it. The AEC-funded scientists elaborated on the ecosystem concept to a new level by making it operational in two ways. In one approach they did this by developing, refining and using methodologies for measuring ecosystem function. Primary production was measured by CO_2 uptake in closed chambers in terrestrial systems and with carbon-14 uptake in aquatic systems. Decomposition was measured with litter bags and respiration was measured with CO_2 evolution from soils. Radioactive tracers were used to identify transfer pathways and to measure rates of movement along pathways in mineral cycles such as at the Cesium Forest at Oak Ridge. Food chain studies tracked energy flow in systems ranging from the Alaskan tundra to the Georgia old fields. Trophic pyramids constructed from the coral reef at Eniwetok demonstrated the role of the Second Law of Thermodynamics in structuring the movements of energy flows. In another approach they created systems diagrams of ecosystems to help understand and communicate the movements of radionuclides and effects of radiation from bomb detonation. These diagrams, some of which are reproduced in this text, helped to visualize the ecosystem as a network or road map of compartments and pathways that connect the compartments into a functional unit. Most of the diagrams illustrated pathways along which materials, such as radionuclides, or energy moved. Other systems models illustrated interactions, rather than flows, such as the sea otter causal diagram from the Aleutian Islands, in Alaska (see Chapter 5). These diagrams eventually became the foundations for mathematical models used in computer simulations (Kwa 1993).

By the late 1960s the radioecologists had produced an elaborate and sophisticated version of the ecosystem concept that was passed on to the International Biological Program (IBP) in which more and better measurements and models were made. The IBP was a massive international program "devoted to understanding the biological basis of human welfare" (Deevey 1968). The U.S. contingent of the IBP chose to focus on the analysis of ecosystems and they developed a huge body of ecological information. As noted by Botkin (1977) concerning the IBP, "The United States provided the largest single effort, involving seven years' work, 1800 scientists, and $43 million in federal grants, divided among five largely separate and individualistic 'biome projects' (desert, tundra, grassland, coniferous and eastern deciduous forest)". Although it ended up to be somewhat controversial, the IBP did much to advance the field of Ecology (see, for example, Reichle 1975) and its heritage has been widely discussed. The irony is that it was the Cold War–era radioecologists who did the fieldwork and the abstraction necessary to operationalize the ecosystem long before the IBP, but this contribution has largely been forgotten. The slight is obvious in the texts on the history of the ecosystem concept by Frank Golley (1993) and David Coleman (2010). In essence, Golley's text covers the ecosystem concept up to 1975 and Coleman's text "picks up where Golley leaves off" (Coleman 2010) with coverage up to the 2000s. It is somewhat amazing how few pages both authors devote to the contribution of Cold War–era Radioecology to the development of the ecosystem concept relative to how much they present on the contributions of the IBP. They certainly acknowledge the work done in the Radioecology projects, but they focus to a greater degree on the role of the IBP research on ecosystems. This slight is even more dramatic because both authors were veterans of Cold War–era Radioecology and because in another publication, Golley (2001) states that "it is probably accurate to say that ecosystem science evolved in the US because of AEC support"!

The refinement and continued development of the ecosystem concept by the Cold War–era radioecologists was a significant accomplishment. The ecosystem is a whole level of the hierarchy of life with properties independent of the lower levels, such as the population or the individual organism. Issues of holism versus reductionism and the idea that the whole (in this case the ecosystem) is greater than the sum of the parts (e.g., populations composed of individual organisms) began to be discussed by Cold War–era radioecologists as they made their measurements and drew their systems diagrams.

A new "invisible college" of Radioecology has emerged, as described in Chapter 10, and there is some indication that the ecosystem concept is coming up again as a useful perspective (Rhodes et al. 2020, Bradshaw et al. 2014). The new generation of radioecologists utilizes the ecosystem created by the first generation who had worked hard to develop it into an operational concept. To some extent they are focusing on lower levels of organization with new tools from molecular biology and genomics that weren't available to the first generation, but there is a continuing need to understand radiation effects at the ecosystem level. Perhaps it is time for a synthesis of the old Cold War–era Radioecology and for a reminder about the utility of the ecosystem concept.

Benchmarks From the Old Radioecology

At the core of the Cold War–era Radioecology research were three interrelated topics: radiosensitivity, radioresilience, and zonation of radiation effects around a source. Each of these topics deals with biodiversity in the context of the ecosystem within which the biodiversity is embedded. The first question that radioecologists asked was: how sensitive are individual species to radiation? This work was simply done by exposing organisms to different amounts of radiation under controlled conditions and recording when they died. The results were typically reported as the lethal dose to kill 50% of the experimental population or LD_{50}. This is a standard approach used as a first step to understand stress effects of any type of toxic pollution. The LD_{50} is a foundation of the field of Ecotoxicology, which evolved in the 1980s to unify pollution studies (Newman 1998). These studies were complicated by the nature of radiation exposure, which could be internal or external to the organism. Because of this issue, much attention was given to dosimetry in all of the early Radioecology projects. Arnold Sparrow put forth the target theory (see Chapter 8) to predict radiosensitivity of a particular species based on the size of the nucleus as a first step, and his theory was tested in a number of the AEC-funded Radioecology projects. As an example, Odum (1970a) reviewed this theory and variations of it in the work at El Verde. Lauren Donaldson and the University of Washington Radioecology group at the Applied Fisheries Lab produced the first synthesis of radiosensitivity across phylogenetic groups of aquatic animals (see Chapter 3). This iconic graph shows relative radiosensitivity is roughly proportional to phylogenetic position with "primitive", simple organisms, such as Protozoa, having low sensitivity and "advanced", complex organisms, such as fishes, having higher sensitivity. Thus, as life gets more complex in terms of number of cell types, organism size and sophistication of organ systems, it becomes more susceptible to damage by radiation. The Applied Fisheries Lab's plot of radiosensitivity was enlarged several times, as in the graph of Whicker and Schultz (1982) that includes 13 organism groups, ranging from virus to mammals. Although this is a concise and revealing way of relating life and biodiversity to radiation, the view is of taxa in isolation, not as part of a systems of interconnected species with network phenomena including indirect and cascading effects. Organisms exist only within ecosystems and as a part of a network, so single species information on radiosensitivity is of limited value in understanding ecological effects of radiation. Radioecology studies of radiosensitivity were done to try to predict which species would survive radiation exposure, but the complexity of the ecosystem produced surprise and unpredictability. Radioresistance is the opposite of radiosensitivity and it occurs within a species due to chance preadaptation to radiation stress. Occasionally, species with relative radioresistance emerged when an ecosystem was experimentally irradiated in the field or in microcosms. Radioresistance is an advantage to a species when resources become available that would otherwise be utilized by a superior competitor that is more sensitive to radiation. This situation occurred, for example, in an irradiated forest in which canopy trees were killed by the radiation, allowing more sunlight to reach the forest floor to support a set of radioresistant understory plant species.

A second major phenomena that was observed in the Cold War–era radiation experiments was that some species exhibited resilience and quickly restored the affected sites once the radiation stress was removed. In some cases these species were a part of the ecosystem before it was stressed, and in other cases these species invaded from surrounding habitats. In particular, a number of studies were carried out on the recovery of the vegetation of an irradiated ecosystem and plant life often returned quickly after the stress stopped. Sometimes this was by regrowth from stump sprouts of damaged plants, but in other cases new species invaded through dispersal from outside. George Woodwell called the invaders "adventive species" in the Brookhaven experiment (see Chapter 8) but the phenomenon was also observed at other sites as well. Another example occurred at the Bikini Atoll where evidence has been found of restoration of the coral reef ecosystem after nuclear bomb testing ended there (Richards et al. 2008). Radiation certainly could destroy ecosystems, but they always would come back in order to utilize the energy sources of the site, thereby restoring structure and function. Radioresilient species may also have been radioresistant, if some residual radiation remained at the site.

The third major finding of Cold War–era Radioecology was the zonation of species around radiation sources: either around a site where an atomic bomb had been detonated or around a radiation source that was experimentally installed in an ecosystem. Zonation is a common, spatial pattern in Ecology in which each species occupies a finite and distinct range of conditions along an environmental gradient. This spatial pattern occurs because no single species can be adapted to the full range of the environmental gradient. Examples are zonation up a mountain where temperature forms a gradient, in an estuary along a salinity gradient or in the intertidal zone along the gradient of tidal inundation. Zonation also was found where radiation intensity formed a gradient outward from a source, such as in a forest where experimental irradiation was studied. This zonation most often took the form of sequential bands of vegetation where "primitive", lower plants such as lichens were found closest to the source, followed sequentially by herbaceous species, shrubs and trees forming bands outward from the source. To a large extent the pattern of zonation around a radiation source was a recapitulation of radiosensitivity, with the most radioresistant species closest to the source and the most radiosensitive species farthest from the source. Radiation "dissects" an ecosystem, as described by Woodwell at Brookhaven (see Chapter 8), by removing species in proportion to exposure intensity. For example, irradiation of a forest unfolds the successional sequence of biodiversity that is stored within the ecosystem along the radiation gradient. It reveals a systems-level "genome" that is "programmed" to regrow the ecosystem after disturbance. The ecosystem contains a large number of species and selection from a stressor such as radiation sorts or filters through this diversity to arrive at combinations that can exist, grow and reproduce under the prevailing environmental conditions. H. T. Odum (1970b) suggested the analogy of the ecosystem as a computer program in the El Verde Project in reference to the repair mechanisms that occurred after irradiation of the rain forest. This idea was supported by the paired studies of irradiation and succession carried out at the Savannah River Lab. There the irradiation experiments (Monk 1966, McCormick 1969) done in several communities essentially moved the systems backwards in time to earlier successional stages that had been outlined independently in research on the entire successional sequence that occurs from bare ground to the hardwood forest climax in the southeastern Coastal Plain (Golley 1974, Johnston and Odum 1956).

Radiation is a dissection tool, like a scalpel, applied to the metaphorical "superorganism" of the ecosystem, and each zone along the radiation gradient is the result of a slice of the radiation scalpel through the ecosystem. The slice is done with energy stress of radiation. In another sense each slice or zone is the solution to an energy balance equation with positive, constructive energies of sunlight, moisture and nutrients compared to the negative, destructive energy of radiation. The solution to this energy balance equation is the species diversity which can exist and live in each zone. Closest to the radiation source, the net balance is negative because there is more destructive radiation than productive energies and no species are able to exist under this condition. But as you move away from the source, the energy balance becomes positive and species are able to exist. In terms of vegetation, the first solution to the energy equation comprises lower plants like lichens and mosses, followed by herbaceous species, shrubs and finally trees as you move outward and away from the radiation source. To continue this algorithmic metaphor, the ecosystem is like an existential computer (Conrad 1995) that solves the energy equation by sorting (to use Woodwell's term) through its species pool for the appropriate biodiversity that can utilize the net energy available at each zone along the radiation gradient. The AEC-funded, Cold War–era radioecologists ran this existential computer in different biomes and recorded the "print-out" in terms of the species composition and the ecosystem function found along the radiation gradient. The initial conditions were different at each study site: from the desert at the NTS, through the short-grass prairie near Colorado State University and to the forests at Brookhaven, Wisconsin, SREL and El Verde, but the form of the "print-out" was the same in each case as the slices of the original ecosystem arranged in a sequential pattern outward from the source. Thus, the recurrence of zonation patterns found in the

Cold War–era Radioecology projects is a powerful empirical finding that reveals what an ecosystem is and how it is put together (Woodwell 1967).

Furthermore, zonation of many properties of ecosystem structure and function was found around radiation sources and the pattern was not always linear. Hump-shaped distributions were sometimes found with maxima at intermediate radiation exposure such as the distribution of litter production in the Brookhaven Forest (see Chapter 8). These kinds of cases suggest stimulatory effects of radiation are possible under certain limited conditions. Another complication of zonation occurred with cases of shielding of radiation by trees, rocks or topography. Shielding provided protection along the radiation gradient in which organisms could survive. Woodwell called these refugia "green shadows" in the Brookhaven Forest in reference to the green plants surviving behind tree trunks in otherwise dead zones (see Chapter 8). Examples of shielding in the irradiation experiments shows that life finds a way to survive by taking advantage of refugia even under conditions of devastating irradiation.

Relevance to the Present Day

The nuclear world created out of the Manhattan Project during World War II and the Cold War that followed is absurd from several perspectives. We have created enough nuclear weapons to destroy humanity many times over and we continue to generate wastes as a by-product of nuclear technologies without a disposal solution that is politically acceptable or even technically sustainable. The Cold War–era Radioecologists, who are the subject of this book, basically did nothing that could have improved these issues. The problem with nuclear wastes was recognized early (Pritchard and Joseph 1963) but the ecologists lacked the skills to deal with the wastes that required secure storage in one way or another until radioactive decay would render them safe. This storage solution required engineering know-how, not ecology. It is somewhat ironic that the radiotracers developed from nuclear technology helped ecologists to confirm and quantify how minerals cycled in Nature but this knowledge did not help them find methods for recycling the radionuclides, which remain locked in storage facilities.

In terms of the prospects of nuclear war, the Cold War–era radioecologists said little about it. Theirs was definitely not a "subversive science" like the Ecology of Shepard and McKinley (1969) or Sears (1964) who advocated a stronger moral stance in regard to environmental problems. As a whole, the radioecologists did not necessarily oppose the Cold War, unlike Carl Sagan's stance against the possibility of nuclear winter (see Chapter 10). As noted by Golley (1993):

> At this time, before the rapid escalation of the Vietnam War, many American ecologists were unconcerned that their studies were closely linked to the military activities. Rather, they tended to accept the cold war as a fact of life and welcomed the opportunities military research made available.

He also goes further and states that "ecologists benefited from the cold war and the Korean War as they studied the base-line ecology of nuclear weapons facilities, the effects of radiation on organisms, and the movement of radioactive materials through food chains" and that the radioecologists "hid within their ivory towers" in regard to the moral issues of the Cold War (Golley 1993). In retrospect, these sound like confessions from a veteran Cold War–era radioecologist and in recent times, social scientists and historians have been critical of various aspects of the AEC-funded Radioecology research (Martin 2018, Raby 2019, Rothschild 2013). For example, DeLoughrey (2012) suggests that "from our contemporary viewpoint, the ecologists' obliviousness to the militarization of their research may seem incredulous." The moral judgment of Cold War–era radioecologists may well deserve more reflective consideration, but that is beyond the scope of this text. They seldom speculated about the large-scale consequences of nuclear war (though see Wolfe 1959, Woodwell 1965) but rather focused in a limited way on particular experiments and data. Their main purpose was to create a body of knowledge about the impacts of radiation that might occur during nuclear war and they did this by developing the ecosystem concept and using it. Perhaps the Cold War–era radioecologists agreed with Lawrence Slobodkin (1995) who stated, "Ecology at one level is as amoral as mathematics".

Since the end of the twentieth century and into the twenty-first century there has been extensive dialogue about the philosophical meaning of the ecosystem concept. Much of this discourse has been published in the journal *Ecosystems*, which began in 1998 (see, for example, Pickett and Cadenasso 2002, Gignoux et al. 2011) and in the *Bulletin of the Ecological Society of America*. Although the ecosystem was posed as the basic unit in Ecology during the 1950s (Evans 1956), some debate later revolved around the question of whether ecosystems even exist! For example, in one exchange the feedback network structure of ecosystems was questioned (Engelberg and Boyarsky 1979), eliciting multiple responses in defense of the concept (Patten and Odum 1981, Knight and Swaney 1981, Jordan 1981). De Laplante (2005) has reviewed the many meanings that have emerged around the ecosystem concept including the ecosystem as object, theory and method, and a pluralistic view is now favored. It is interesting to read the ecological literature from the 1950s and 1960s and to get a sense of the simpler, perhaps more intuitive though naïve sense of the ecosystem from that time. The Cold War–era radioecologists didn't have the time or weren't ready for philosophical debate about the ecosystem concept, which would come later. There was an urgency to comprehend how Nature responds to the novel stress of radiation because nuclear war was imminent. They took the concept from Tansley and Lindeman and used it to help reduce the uncertainties about the ecological consequences of nuclear war. It seems we have more of the luxury of indulging in philosophy now than the Cold War–era radioecologists had. Maybe we need to feel a sense of urgency again in regard to the terror of nuclear disasters like Chernobyl and the dilemma of what to do with radioactive waste that continues to accumulate from nuclear technologies, which are not going away. Arsenals of nuclear weapons are rising again due to unrest in Iran and North Korea and with the threat of a nuclear attack in the war between Ukraine and Russia. Also, the fear of global climate change from the burning of fossil fuels has renewed interest in nuclear power as an alternative with a smaller carbon footprint, so more reactors may be soon be built. These developments in the post–Cold War age mean that there is a continuing need to understand the impacts on ecosystems from radiation.

In the Cold War–era Radioecology projects funded by the AEC, stress from radioactivity was found to simplify ecosystems and to reduce structure and function. The experimental work was limited in spatial scale to a single bomb detonation or to a single radiation source inserted into an ecosystem, but the results were consistent. Ecosystems were rarely eliminated over any sizable area or for any significant length of time. Radioecologists' field and laboratory studies demonstrated that life "finds a way" to survive and to restore itself in every one of the AEC-funded studies that spanned most of the Earth's biomes. A number of surprising examples of radioresistance and radioresilience were found, demonstrating how ecosystems can self-organize through selection of species so that energy sources continue to be exploited even in the face of devastating stress (Table 11.1). These species emerged from the larger pool of biodiversity once the ecosystem was irradiated and the diversity was reduced. Most were minor elements of the systems before irradiation and they survived due to chance preadaptation, since they did not evolve with the high levels of radiation exposure. In a sense, these kinds of species demonstrate the notion that the whole is more than the sum of the parts because they allow the ecosystem (the whole) to continue functioning, even though the majority of the species (the parts) were eliminated.

Today there is much discussion of "novel" ecosystems that emerge through self-organization under new conditions of pollution and environmental impact (Hobbs et al. 2013, Martinuzzi et al. 2015; Radeloff et al. 2015), and the AEC-sponsored Radioecology research of the 1950s and 1960s would fit nicely into this topic area. Novel ecosystems are composed of new combinations of species that arise from global climate change or the introduction of exotic species. Examination of the old literature on Cold War–era radiation experiments may lend insight into structure and function of these kinds of systems. Mutations occasionally were found to occur in the radiation experiments, but this was rare and no monsters ever emerged from the AEC-funded projects. The ecosystem was found to be more resistant to radiation than any individual species because of the ecological substitutions of species that occurred under the selection pressures of radiation. An example of this condition was seen with the forest floor microcosms at El Verde (see Chapter 7). Replicate microcosms were placed in the rain forest plot that was exposed to radiation and in the control plot with no radiation. No differences were found in the metabolism of the systems after the three-month irradiation period, which was attributed to substitutions in the microbial populations of the microcosms (Odum and Lugo 1970). The ecosystem is certainly more resistant than humans, who are probably the most radiosensitive

TABLE 11.1 Listing of species found at the AEC-funded Radioecology projects that had special qualities of radioresistance or radioresilience. These species emerged at the sites through self-organization of a community that had been irradiated.

Species	*Location*	*Special Feature*
Roof rat (*Rattus rattus*)	PPG	Survived hydrogen bomb detonation
Tumbleweed (*Salsola* spp.)	NTS	First to colonize because of tumbleweed habit of dispersal
Mentzeliz albica, a desert annual	NTS	Found in higher abundance in bombed areas due to greater soil moisture once desert pavement is broken
Freshwater clams (Unionidae)	ORNL	Deposited Sr-90 in their shells and acted as biomonitors
Chironomus tetans, aquatic fly larvae	ORNL	Survived in mud of White Oak Lake under high radiation exposure
Palicourea riparia, a small understory tree	El Verde	Survived high radiation intensity
Digitaria spp. (crabgrass)	BNL	Prostrate growth form helped it grow under high radiation exposure in the old field
Carex pennsylvanica, a sedge	BNL	Vegetative growth and low, graminoid form helped it grow under high radiation exposure in the forest
Schizothrix calcicola, blue-green alga	SREL microcosm	Increased growth once a radiosensitive competitor was inhibited by radiation exposure
Fimbristylis autumnalis, a sedge	SREL	Grew in high radiation exposure zone of an old field but was not found in the community before irradiation

species on the planet. The empirical findings from the Cold War–era Radioecology projects might not hold up under the hypothetical conditions of large-scale, all-out nuclear war, as laid out in the nuclear winter studies (Grover and Harwell 1985), but they stand as a credible body of research about the effects of artificial radiation on life. These findings have been largely forgotten in the history of ecology and are worth revisiting.

The Anthropocene, the geologic era of human dominance of the Earth, is sometimes said to have begun in 1945 with the detonation of the first atomic bomb in the desert of southern New Mexico. There is good reason for this timing, since the radionuclides released from bomb testing formed a marker strata of radioactivity in the sediments of lakes and coastal marshes around the globe. In an analogous way, Cold War–era radioecologists left a marker in the ecological literature with the publications of their research. Their stories mark the time when the ecosystem concept became operational and the results of their research demonstrated the vitality of life at the ecosystem level of organization. In the Cold

War–era experiments, radiation stress changed the species composition of the ecosystem and reduced diversity, but a simpler system emerged to continue processing sunlight energy and to continue cycling minerals, thereby demonstrating the resistance of the ecosystem. After the stress was removed, diversity returned, slowly but surely, to increase ecosystem function, demonstrating the resilience of the system. These findings provide a faith in the ecosystem to survive even under the most extreme form of anthropogenic pollution, artificial radiation.

Literature Cited

Adams, L. W., G. C. White and T. J. Peterle. 1976. Tritium kinetics in a freshwater marsh. Pp. 96–103. In: C. E. Cushing, Jr. (ed.). Radioecology and Energy Resources: Proceedings of the Fourth National Symposium on Radioecology. Dowden Hutchinson and Ross, Stroudsburg, PA.

Alexakhin, R. M. 2006. Radioecology: History and state-of-the-art at the beginning of the 21st century. Pp. 159–168. In: A. A. Cigna and M. Durante (eds.). Radiation Risk Estimates in Normal and Emergency Situations. Springer, Dordrecht, The Netherlands.

Becker, C. C. 1990. Aquatic Bioenvironmental Studies: The Hanford Experience 1944–1984. Elsevier, Amsterdam.

Botkin, D. B. 1977. Bits, bytes, and IBP. Bioscience 27:385.

Bradshaw, C., L. Kapustka, L. Barnhouse, J. Brown, P. Ciffroy, V. Forbes, S. Geras'kin, U. Kautsky and F. Brĕchignac. 2014. Using an ecosystem approach to complement protection schemes based on organism-level endpoints. Journal of Environmental Radioactivity 136:98–104.

Coleman, D. C. 2010. Big Ecology: The Emergence of Ecosystem Science. University of California Press, Berkeley, CA.

Conrad, M. 1995. The ecosystem as an existential computer. Pp. 609–622. In: B. C. Patten and S. E. Jorgensen (eds.). Complex Ecology: The Part-Whole Relation in Ecosystems. Prentice Hall, Englewood Cliffs, NJ.

Daniel, C. P. and R. B. Platt. 1968. Direct and indirect effects of short term ionizing radiation on old-field succession. Ecological Monographs 38:1–30.

Davis, J. J. and R. F. Foster. 1958. Bioaccumulation of radioisotopes through aquatic food chains. Ecology 39:530–535.

Deevey, E. S. 1968. Why an international biological program? Bioscience 18:761.

de Laplante, K. 2005. Is ecosystem management a postmodern science? Pp. 397–416. In: K. Cuddington and B. E. Beisner (eds.). Ecological Paradigms Lost, Routes of Theory Change. Elsevier, Amsterdam.

DeLoughrey, E. M. 2012. The myth of isolates: Ecosystem ecologies in the nuclear Pacific. Cultural Geographies 20:167–184.

Duke, T. W., J. N. Willis and T. J. Price. 1966. Cycling of trace elements in the estuarine environment. I. Movement and distribution of Zinc 65 and stable Zinc in experimental ponds. Chesapeake Science 7:1–10.

Engelberg, J. and L. L. Boyarsky. 1979. The noncybernetic nature of ecosystems. American Naturalist 114:317–324.

Evans, F. C. 1956. Ecosystem as the basic unit in ecology. Science 123:1127–1128.

Fraley, L. and F. W. Whicker. 1971. Response of a native shortgrass plant stand to ionizing radiation. Pp. 999–1006. In: D. J. Nelson (ed.). Radionuclides in Ecosystems,

Proceedings of the Third National Symposium on Radioecology. CONF-710501-P2, U.S. Atomic Energy Commission, Washington, DC.

Gignoux, J., I. D. Davies, S. R. Flint and J.-D. Zucker. 2011. The ecosystem in practice: Interest and problems of an old definition for constructing ecological models. Ecosystems 14:1039–1054.

Golley, F. B. 1974. Structural and functional properties as they influence ecosystem stability. Pp. 97–102. In: Proceedings of the First International Congress of Ecology: The Structure, Functioning and Management of Ecosystems. The Hague, The Netherlands.

Golley, F. B. 1993. A History of the Ecosystem Concept in Ecology. Yale University Press, New Haven, CT.

Golley, F. B. 2001. Establishing the network. Pp. 38–68. In: G. W. Barrett and T. L. Barrett (eds.). Holistic Science: The Evolution of the Georgia Institute of Ecology (1940–2000). Taylor & Francis, New York, NY.

Grover, H. D. and M. A. Harwell. 1985. Biological effects of nuclear war II: Impact on the biosphere. Bioscience 35:576–583.

Hobbs, R. J., E. S. Higgs and C. M. Hall (eds.). 2013. Novel Ecosystems, Intervening in the New Ecological World Order. John Wiley & Sons, Chichester.

Johnson, W. C. and N. R. French. 1981. Soviet union. Pp. 343–383. In: E. J. Kormondy and J. F. McCormick (eds.). Handbook of Contemporary Developments in World Ecology. Greenwood Press, Westport, CT.

Johnston, D. W. and E. P. Odum. 1956. Breeding bird populations in relation to plant succession on the Piedmont of Georgia. Ecology 37:50–62.

Jordan, C. F. 1981. Do ecosystems exist? American Naturalist 118:284–287.

Klechkovskii, V. M., G. G. Polikarpov and R. M. Aleksakhin (eds.). 1973. Radioecology, Translated from Russian by N. Kaner and H. Mills. John Wiley & Sons, New York, NY.

Klement, A. W., Jr., C. F. Lytle and V. Schultz. 1968. Russian Radioecology: A Bibliography of Soviet Publications with Citations of English Translations and Abstracts. TID-5915, U.S. Atomic Energy Commission, Washington, DC.

Knight, R. L. and D. P. Swaney. 1981. In defense of ecosystems. American Naturalist 117:991–992.

Kwa, C. 1993. Radiation ecology, systems ecology and the management of the environment. Pp. 213–249. In: M. Shortland (ed.). Science and Nature, Essays in the History of the Environmental Sciences. British Society for the History of Science, London, UK.

Martin, L. J. 2018. Proving grounds: Ecological fieldwork in the Pacific and the materialization of ecosystems. Environmental History 23:567–592.

Martinuzzi, S., G. I. Gavier-Pizarro, A. E. Lugo and V. C. Radeloff. 2015. Future land-use changes and the potential for novelty in ecosystems of the United States. Ecosystems 18:1332–1342.

McCormick, J. F. 1969. Effects of ionizing radiation on a pine forest. Pp. 74–87. In: D. J. Nelson and F. C. Evans (eds.). Symposium on Radioecology, Proceedings of the Second National Symposium. CONF-670503, U.S. Atomic Energy Commission, Washington, DC.

McCormick, J. F. and R. B. Platt. 1962. Effects of ionizing radiation on a natural plant community. Radiation Botany 2:161–188.

Monk, C. D. 1966. Effects of short-term gamma irradiation on an old field. Radiation Botany 6:329–335.

Nelson, D. J. (ed.). 1971. Radionuclides in Ecosystems, Proceedings of the Third National Symposium on Radioecology. CONF-710501, U.S. Atomic Energy Commission, Washington, DC.
Nelson, D. J. and F. C. Evans (eds.). 1969. Symposium on Radioecology, Proceedings of the Second National Symposium. CONF-670503, U.S. Atomic Energy Commission, Washington, DC.
Newman, M. C. 1998. Fundamentals of Ecotoxicology. Ann Arbor Press, Chelsea, MI.
Odum, H. T. 1970a. Microscopic order in the forest. Pp. G-3–G-14. In: H. T. Odum and R. F. Pigeon (eds.). A Tropical Rain Forest. U.S. Atomic Energy Commission, Washington, DC.
Odum, H. T. 1970b. Summary, an emerging view of the ecological system at El Verde. Pp. I-191–I-289. In: H. T. Odum and R. F. Pigeon (eds.). A Tropical Rain Forest, U.S. Atomic Energy Commission, Washington, DC.
Odum, H. T. and A. E. Lugo. 1970. Metabolism of forest-floor microcosms. Pp. I-35–I-56. In: H. T. Odum and R. F. Pigeon (eds.). A Tropical Rain Forest. U.S. Atomic Energy Commission, Washington, DC.
Osburn, W. S. 1967. Ecological concentration of nuclear fallout in a Colorado Mountain watershed. Pp. 675–709. In: B. Åberg and F. P. Hungate (eds.). Radioecological Concentration Processes. Pergamon Press, Oxford, UK.
Patten, B. C. and E. P. Odum. 1981. The cybernetic nature of ecosystems. American Naturalist 118:886–895.
Pickett, S. T. A. and M. L. Cadenasso. 2002. The ecosystem as a multidimensional concept: Meaning, model, and metaphor. Ecosystems 5:1–10.
Platt, R. B. 1965. Radiation effects on plant populations and communities: Research status and potential. Health Physics 11:1601–1606.
Platt, R. B. and J. F. McCormick. 1964. Manipulatable terrestrial ecosystems. Ecology 45:649–650.
Pritchard, D. W. and A. B. Joseph. 1963. Disposal of radioactive wastes: Its history, status, and possible impact on the environment. Pp. 27–31. In: V. Schultz and A. W. Klement, Jr. (eds.). Radioecology, Proceedings of the First National Symposium on Radioecology. Reinhold Publishing Corporation, New York, NY.
Pruter, A. T. and D. L. Alverson (eds.). 1972. The Columbia River Estuary and Adjacent Ocean Waters, Bioenvironmental Studies. University of Washington Press, Seattle, WA.
Raby, M. 2019. "Slash-and-burn ecology": Field science as land use. History of Science 57:441–468.
Radeloff, V. C. and 19 co-authors. 2015. The rise of novelty in ecosystems. Ecological Applications 25:2051–2068.
Reichle, D. E. 1975. Advances in ecosystem analysis. Bioscience 25:257–264.
Rhodes, O. E., Jr. and 51 co-authors. 2020. Integration of ecosystem science into radioecology: A consensus perspective. Science of the Total Environment 740(2020):140031.
Rice, T. R. and T. W. Duke. 1968. Research Facilities of the Radiobiological Laboratory, Bureau of Commercial Fisheries. Circular 298, U.S. Department of the Interior, U.S. Fish and Wildlife Service, Washington, DC.
Richards, Z. T., M. Beger, S. Pinca and C. C. Wallace. 2008. Bikini Atoll coral biodiversity resilience five decades after nuclear testing. Marine Pollution Bulletin 56:503–515.
Rothschild, R. 2013. Environmental awareness in the Atomic Age: Radioecologists and nuclear technology. Historical Studies in the Natural Sciences 43:492–530.
Schultz, V. and A. W. Klement, Jr. (eds.). 1963. Radioecology, Proceedings of the First National Symposium on Radioecology. Reinhold Publishing Corporation, New York, NY.

Sears, P. B. 1964. Ecology: A subversive subject. Bioscience 14:11.
Shepard, P. and D. McKinley (eds.). 1969. The Subversive Science, Essays Toward an Ecology of Man. Houghton Mifflin Co., Boston, MA.
Slobodkin, L. B. 1995. Linking species and ecosystems through training of students. Pp. 306–312. In: C. G. Jones and J. H. Lawton (eds.). Linking Species & Ecosystems. Chapman & Hall, New York, NY.
Whicker, F. W. and V. Schultz. 1982. Radioecology: Nuclear Energy and the Environment, Volume 2. CRC Press, Boca Raton, FL.
White, G. C., L. W. Adams and T. A. Bookhout. 1978. Simulation model of tritium kinetics in a freshwater marsh. Health Physics 34:45–54.
Whittaker, R. H. 1961. Experiments with radiophosphorus tracer in aquarium microcosms. Ecological Monographs 31:157–188.
Wolfe, J. N. 1959. Long-Time Ecological Effects of Nuclear War. TID-5561, U.S. Atomic Energy Commission, Washington, DC.
Woodwell, G. M. (ed.). 1965. Ecological Effects of Nuclear War. BNL 917 (C-43), Brookhaven National Laboratory, Upton, NY.
Woodwell, G. M. 1967. Radiation and patterns of nature. Science 156:461–470.

INDEX

For Product Safety Concerns and Information please contact our EU
representative GPSR@taylorandfrancis.com
Taylor & Francis Verlag GmbH, Kaufingerstraße 24, 80331 München, Germany

www.ingramcontent.com/pod-product-compliance
Ingram Content Group UK Ltd.
Pitfield, Milton Keynes, MK11 3LW, UK
UKHW020141200926
475702UK00005B/859
9781032257631